D0148406

CAMBRIDGE MONOGRAPHS ON PHYSICS

GENERAL EDITORS

A. HERZENBERG, PH.D.

Reader in Theoretical Physics, University of Manchester

M. M. WOOLFSON, D.SC.

Professor of Theoretical Physics, University of York

J. M. ZIMAN, D.PHIL., F.R.S.

Professor of Theoretical Physics, University of Bristol

RADIOTELESCOPES

RADIOTELESCOPES

BY

W. N. CHRISTIANSEN

Professor of Electrical Engineering
University of Sydney

AND

J. A. HÖGBOM

Senior Astronomer
Netherlands Foundation for
Radio-astronomy

CAMBRIDGE

AT THE UNIVERSITY PRESS

1969

Published by the Syndics of the Cambridge University Press
Bentley House, 200 Euston Road, London N.W.1
American Branch: 32 East 57th Street, New York, N.Y.10022

© Cambridge University Press 1969

Library of Congress Catalogue Card Number: 69–16279
Standard Book Number: 521 7054 6

Printed in Great Britain
at the University Printing House, Cambridge
(Brooke Crutchley, University Printer)

CONTENTS

CHAPTER 5

Some more theory

CHAPTER 6

Unfilled-aperture antennas

CHAPTER 7

Aperture synthesis

CHAPTER 8

Sensitivity

PREFACE

The first radio astronomical observations were made with an antenna designed for short-wave radio communication. During the early years of radio astronomy radar-type antennas were used with beam widths varying between about 1° at centimetre wavelengths and 10° at longer wavelengths. These soon failed to satisfy radio-astronomers.

In the late 1940s and early 1950s many new types of radio-telescope were invented. Some of the most important were based on the optical interferometer. The first of these used reflection from the surface of the sea to provide the second ray path but later instruments had a pair of antennas with variable spacing. Out of such interferometers developed the powerful technique of 'aperture synthesis' which forms the basis of many new telescopes and which is suited to observations of unchanging sources of radiation. Simultaneously there were invented new radiotelescopes which were not so limited in their application. The best known of these are the one-dimensional grating and the compound interferometer and the two-dimensional cross antenna. At this time first use was made of the earth's rotation to synthesize a two-dimensional antenna from a one-dimensional antenna. Very long baseline interferometry also developed rapidly. Simultaneously the physical size of the older types of antenna such as the parabolic reflector telescope increased enormously.

The result of this activity has been that radio maps of the sky now have an angular resolution several orders of magnitude better than was possible 20 years ago, and very long baseline interferometers have enabled radio astronomers to determine the angular sizes of objects down to $\frac{1}{1000}$ second of arc. In addition, information is now obtained simultaneously from many points in the sky instead of from only one.

It is unfortunate that the reporting of these developments is either scattered amongst a wide range of journals or else has never been put on paper. The authors of this book have spent a great part of this period in designing radiotelescopes, and we aim to present to newcomers in radio astronomy a short survey of the development of radiotelescopes during the last 20 years, with enough simplified theory to enable them to understand the fundamentals of radiotelescope design. We have tried also to pass on some of the experience we have gained in our design work. The writing of the book has been a spare-time activity of authors separated by a great distance, and if the book lacks the polish and balance associated with the publication of a well-organized lecture course, we hope that the roughest edges have been removed as the result of many trips that the manuscript has made across the world from one writer to the other.

The references are to papers which we have found useful, and can recommend for further reading. Much work covered by later surveys has not been mentioned or else has been mentioned only in part, and we apologise to many workers whose valuable contributions appear anonymously.

We wish to thank a number of radio astronomers who have read parts of the manuscript; at the same time we absolve them from any responsibility for errors. We thank particularly Prof. C. A. Muller, Prof. B. Y. Mills, Dr J. P. Wild, Dr S. F. Smerd, Dr R. F. Mullaly, Mr B. J. Elliott and Dr K. J. Wellington.

We should like to express our gratitude also for the work of those concerned with the preparation of the manuscript—Miss D. E. Wood, Mrs K. Magarey, Mevr Focquin de Grave-Polak, Fraulein Winkelhag and Miss M. Conley. Finally we gratefully acknowledge that this book would not have been completed without the patient organization of E. M. Christiansen.

W.N.C.
J.A.H.

CHAPTER 1

INTRODUCTION TO RADIOTELESCOPES

1.1. The purpose of a radiotelescope

For hundreds of thousands of years men watched the sun, the moon, stars, meteors and comets. These celestial objects were remote; they could not be touched, heard, smelt or tasted. By visual observations alone men slowly learned the laws of movement of these mysterious things. The invention five hundred years ago of the optical telescope, which followed, inevitably though tardily, from the manufacture and use of transparent glass, allowed the celestial bodies to be seen in much greater detail than before. A study of the physical laws which are obeyed by optical devices revealed that a telescope must have an aperture very large compared with the wavelength of light to reveal the very fine structure of the object being observed. During the last century, therefore, telescopes of very large aperture have been built and these, with the spectroscope, have given more knowledge of the external universe than was obtained during the previous hundreds of millennia.

For nearly a hundred years it has been known from theory that the sun and stars emit electromagnetic radiations ranging from short to long wavelengths or from X-rays to radio emissions. At the end of the nineteenth century attempts were made to receive radio emissions from the sun. These were not successful and it was not until radio knowledge and techniques had developed much further that radio waves made their first contribution to astronomy. The new astronomical knowledge was unexpectedly important and radio astronomy rapidly became a major branch of astronomy. The main reason for this is that radio waves are about one million times longer than light waves and regions of space that are opaque to light waves are frequently transparent to radio waves and vice versa. The million-fold difference in the wavelengths, however, makes it extremely difficult to see as much detail in the sky with radiotelescopes as with optical telescopes; to see equal amounts of detail a radiotelescope has to be a million times larger. A radiotelescope receiving emissions at a wavelength of 20 cm, for example, needs an aperture of about 1000 km to give the same amount of detail (resolution) as would be provided by a large optical telescope. The attainment of high resolving power (the ability to distinguish between two close objects) has been one of the main goals of radio-astronomers during the last three decades. This has led to the building of huge mechanical structures which are scaled-up versions of optical devices. In addition, it has led to the invention of new devices which have no optical analogue.

1.2. The objects studied by radio astronomers

Directly observable quantities associated with a celestial object include its angular position with respect to some celestial coordinate system, its angular size and the flux density in a certain range of wavelengths (or frequency) of the radiated power which reaches the telescope. The angular size of objects available for observation by radio astronomers ranges from a lower limit which is imposed by irregularities in the earth's atmosphere to several degrees. The flux density of the radiation from the objects studied ranges also over many orders of magnitude but is always extremely small.

Flux density S is measured in terms of the power falling on unit area of a surface which is normal to the direction of arrival of the radiation. The unit is the watt per square metre per unit frequency bandwidth (in cycles per second, or Hertz) and is written

$$W m^{-2} Hz^{-1}.$$

The flux density is a function of the frequency at which the measurements are made; this function is the *radio spectrum* of the source. Figure 1.1 shows on a logarithmic scale the radio spectra of a number of sources. The range of frequencies in the diagram, 10 MHz ($\lambda = 30$ m) to 30,000 MHz ($\lambda = 1$ cm), includes most of the range that is useful for observations from the ground. This 'radio window' through the atmosphere is limited at the low frequency end by ionospheric reflection and at the high frequency end by molecular absorption bands of water and oxygen in the atmosphere.

The flux density scale illustrates how minute are the energies involved even for the most intense sources, and flux densities as low as 10^{-29} $Wm^{-2}Hz^{-1}$ can be measured with present-day instruments. A large radiotelescope may collect the radiation over a 10 MHz frequency band falling on an area of $10^4 m^2$ and the receiver is capable of measuring a power of $10^{-29+7+4} = 10^{-18}$ W. Radio-

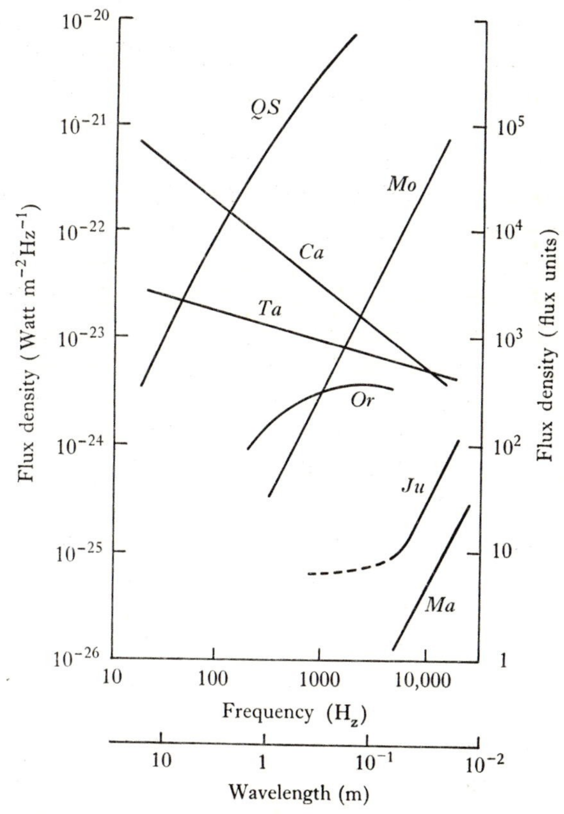

Figure 1.1. The radio spectra of some prominent radio sources. *QS*, quiet Sun; *Mo*, Moon; *Ca*, source in Cassiopeia; *Ta*, source in Taurus (Crab Nebula); *Or*, Orion Nebula (an optically thin source); *Ju*, Jupiter; *Ma*, Mars.

telescopes are among the most sensitive instruments built by man and the measurements are easily rendered useless by interference, particularly man-made transmissions, which often involve powers of more than 10^5 W.

The spectra in Figure 1.1 can be divided into two main groups:

(*a*) *Thermal radiation*. The spectrum of this is given with sufficient accuracy by Rayleigh's classical approximation to the radiation law at the frequencies and intensities usually encountered in radio-astronomy. A 'black-body' radiator at a temperature T °K should, therefore, have a flux density $S \propto \nu^2 T$. The *spectral index* (exponent in the frequency dependence) for a black-body radiator is thus $\alpha = 2$. The spectra of the planets, the Moon and the Sun are of this general type (the Sun and Jupiter have, in addition, a variable non-thermal component), and the corresponding *radio brightness temperatures* are in reasonable agreement with the temperatures of these objects as measured by other methods. The brightness temperature of the Sun changes by a factor of about one hundred from one end of the frequency scale to the other. The high temperature radiation comes from the Sun's outer atmosphere (the corona), which has a temperature of a few million degrees and is opaque to low-frequency radiation. It is transparent to high-frequency radiation, which therefore reaches the Earth from the much cooler (about 10^4 °K) solar surface layers. A region that is partly transparent can itself radiate only a certain fraction of the normal black-body radiation. This fraction is proportional to the opacity (optical depth) which in very 'thin' regions (consisting mainly of ionized hydrogen) is itself proportional to $1/\nu^2$. The black-body radiation is $\propto \nu^2$ and the region will, therefore, have spectral index $\alpha = 0$ (a 'flat' spectrum), where the flux density is independent of the frequency, until towards the low frequencies the region becomes completely opaque and the spectrum curves take up the slope of a black-body radiator. The spectrum of the Orion nebula is an example of this general behaviour, but it is complicated by the presence in the nebula of several regions of different density.

(*b*) *Non-thermal radiation.* There are processes which are much more effective in producing radio emissions than the random collisions between electrons and ions. These processes are lumped under the name of non-thermal emission. Of these, the most common appears to be synchrotron emission (sometimes called magnetic bremsstrahlung) by relativistic electrons accelerated in their passage through magnetic fields. The theory allows a certain range of spectral index and also some curvature in the individual spectra, as is indeed observed for a number of sources. The radiation emitted by the synchrotron mechanism would be expected to show linear polarization and this has been observed in many radio sources, including the general 'background' radiation from our own Galaxy. The sources of non-thermal radiation are amongst the most intense and most interesting to be studied by astronomers. It appears that many of these are at enormous distances, far beyond the reach of optical telescopes, and much effort is going into the study of these sources and their relation to questions of cosmology such as the history and the geometry of the Universe on the largest scale.

Apart from the smooth continuum spectra of Figure 1.1 there are spectral lines which can be studied with special receiving equipment. The best known are those at 1420 MHz ($\lambda = 21$ cm) from the neutral hydrogen atom and a family of lines at about 1667 MHz ($\lambda = 18$ cm) from the hydroxyl radical. Several recombination lines (high n) of hydrogen have also been discovered. The spectral lines are extremely important because, amongst other things, they make it possible to measure radial velocities by their Doppler frequency shift. The general structure and the motions in the Galaxy and in many external galaxies have been mapped by observations of the 21 cm line of neutral hydrogen.

The Sun is the most intense source and also the most variable. There is, in addition to the thermal 'quiet' Sun radiation, a non-thermal component associated with active regions on the Sun. The flux density can increase during a 'burst' of radiation to a level of $10^{-18}\,\mathrm{W m^{-2} Hz^{-1}}$ or more during periods of great activity. It is, however, still much smaller than the normal flux from the Sun at optical frequencies.

The other sources of radio emission in the sky are weaker than the Sun but not by such a large factor as in optical astronomy. The Earth's atmosphere scatters radio waves very much less than it scatters light waves, and some radio measurements may, if necessary, be made in daytime. At the upper wavelength limit of the radio window, at about $\lambda = 20$ m, a large part of the general background radiation from the Galaxy has a radio emission per unit solid angle or *radio brightness* which is comparable with that of the quiet Sun. At a wavelength of 3 m it has dropped to less than 1 per cent of that of the quiet Sun. The background is even less significant at centimetre wavelengths. These changes illustrate the difference between the spectrum of the quiet Sun and that of the background.

Two objects in the northern sky, the well-known radio sources in Cassiopeia and Cygnus, rival at the longer wavelengths the quiet Sun in radio energy flux received at the Earth. Other sources are weaker and the number of sources that can be detected increases rapidly as one goes to weaker sources. If the sources were equal in emitting power and were distributed uniformly in a Euclidean space, then the total number of sources n with a flux density $S \geqslant S_0$ would be proportional to $S_0^{-1.5}$. Because of its significance in cosmology the relation between n and S_0 has engaged the attention of radio-astronomers for more than a decade.

If a source is larger in angular extent than the main response (beam) of the radiotelescope it is said to be *resolved*. What is measured by the radiotelescope in such circumstances is the flux emitted by the parts of the source which lie within the beam of the telescope. This quantity is related to the *radio brightness B* of the particular region of the source and will be discussed in the next chapter.

1.3. Filled-aperture radiotelescopes

1.3.1. *The parabolic reflecting antenna*

The first radiotelescope that was deliberately designed for astronomical purposes was built by Grote Reber[101] in 1937. This was a paraboloid of revolution, the radio analogue of an optical reflecting telescope and a descendant of the first directional radio antenna ever made, that used by Heinrich Hertz in his early experiments on radio waves.

Geometrical optics is of limited usefulness in studying parabolic reflector radiotelescopes. The concepts of rays, straight line propagation, geometrical shadows, etc., serve as approximations to the exact wave propagation equations and these approximations may be used when the scale of the phenomena is much larger than the wavelength. These concepts are very useful in optics because the wavelength of light ($\sim 5 \times 10^{-5}$ cm) is minute compared with most optical apparatus (mirrors, lenses, the human eye, etc.). Special experiments are necessary to show optical diffraction effects. In radio-astronomy, on the other hand, we deal with wavelengths of the same order of size as man-made equipment (millimetres to tens of metres) and 'rays', 'shadows' and similar concepts should be used only for rough descriptions of the mode of operation of radiotelescopes.

The parabolic reflector radiotelescope, therefore, is better described by using Huygen's principle: we should calculate the energy at the focus as the sum of the contributions from all parts of the incoming wave-front, taking into account the different path lengths and therefore different radio-frequency phases at the focus (Figure 1.2). The resultant field distribution in the focal plane of a

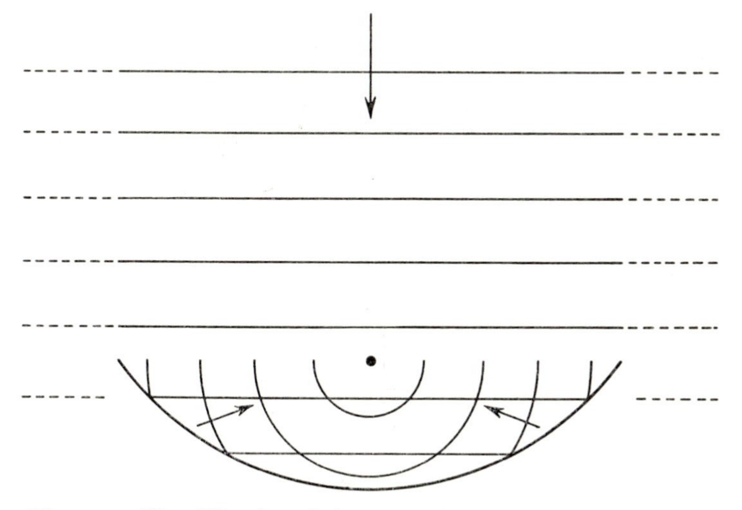

Figure 1.2. The diffraction of plane waves by a paraboloidal reflecting surface and the resultant contracting spherical wave system.

circular-aperture paraboloid is shown in Figure 1.3*a*. The power available for absorption by a 'feed' antenna placed at the focus is proportional to the square of intensity of this field (Figure 1.3*b*).

Figure 1.3*b* shows how the antenna output varies with the position of a source relative to the telescope axis, and it therefore shows how the effective energy absorbing area changes with the direction of incidence of the waves. This function (usually normalized to unity in the maximum direction) is called the *power pattern* or simply the 'beam' of the antenna.

Even in the absence of scintillation by the Earth's atmosphere the light from a star situated along the axis of an ideal paraboloid mirror

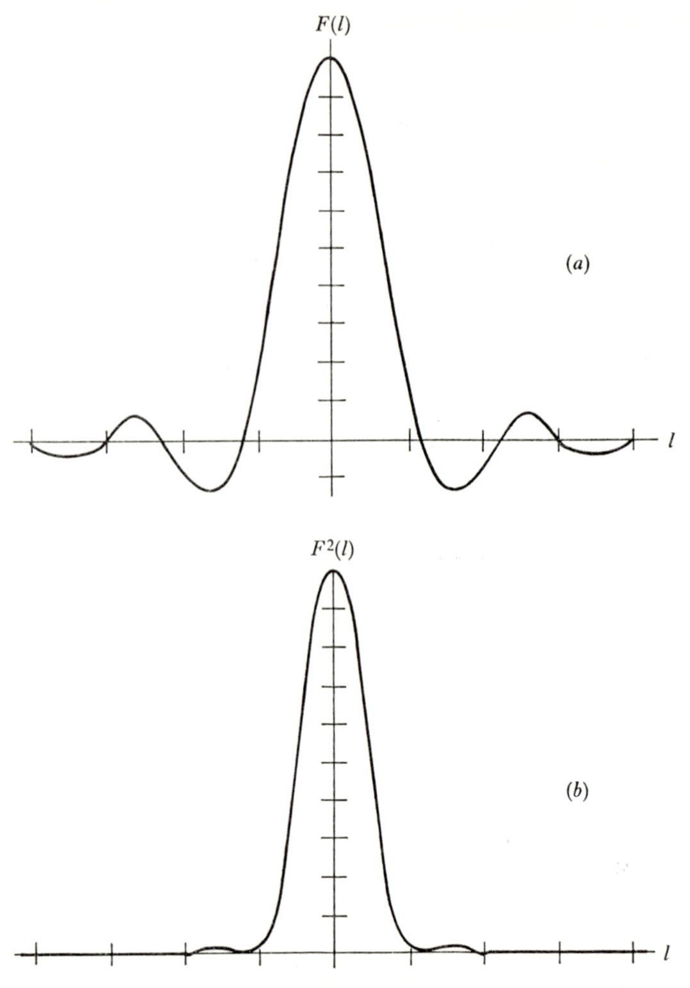

Figure 1.3. (*a*) The field-strength of electromagnetic radiation in the focal plane of a paraboloidal antenna as a function of the distance from the focus.

The curve represents also the voltage appearing at the receiver terminals as a function of the direction of the source of radio emission (with respect to the axis of the paraboloid).

(*b*) The power appearing at the antenna terminals as a function of the direction of the emitting source.

The curve represents also the power distribution in the focal plane of the paraboloid as a function of the distance from the focus (Airy diffraction disk).

is concentrated, not at a point, but into a small yet finite area in the focal plane, the so-called Airy disk. If, in the radiotelescope, we explore the area around the geometrical focus with a small pickup antenna probe, we shall find that the radio-frequency energy from a point source on the telescope axis is distributed approximately according to the formula for the Airy-diffraction disk, well known in optics. If, instead of exploring the focal plane, we fix the probe at the geometrical focus and steer the telescope so that the axial direction moves with respect to the star, we obtain the same Airy disk distribution. The *Airy disk* of optics and the *power pattern* of the antenna are therefore different ways of expressing the same phenomenon.

A photograph taken with an ideal optical reflecting telescope in perfect atmospheric conditions is 'smoothed' or 'blurred' to a certain extent because every detail is spread out over an area corresponding to the Airy-diffraction disk. A radio map obtained, for example, by measuring the power output from the antenna as the telescope beam is scanned across the region will be smoothed, because every measurement corresponds to the average radio flux from a finite solid angle of sky, i.e. that occupied by the beam. Plate 1 (facing p. 96) gives an illustration of how this smoothing affects the measurements. Structure on an angular scale larger than the beam, for example the spiral structure, is clearly visible though the finer details have been lost. The former is said to be resolved by the telescope and the latter unresolved.

The resolving power depends on λ/D and can therefore be improved by using either a shorter wavelength λ or a larger diameter D for the telescope aperture. The first possibility is limited by absorption in the Earth's atmosphere at wavelengths shorter than a few millimetres. It is restricted also because radio-astronomers are interested, in general, in the spectral characteristics of radiation from radio sources and so would not be satisfied with observations at millimetre wavelengths only. The minimum usable wavelength is proportional to the antenna irregularities or structural deflections and these are by no means independent of D, as will be shown in Chapter 3. There is, in fact, a limit to the size of a steerable paraboloid, and even this limit is usually far below the size required for adequate resolving power in radio-astronomy.

1.3.2. *Other reflector-type radiotelescopes*

Because of the inherent limitations in size of steerable paraboloids, attempts have been made to design telescopes in which the reflecting surface is fixed or, if movable, is supported from the ground at many points, so that weight and cost do not increase as rapidly with size as they do with steerable paraboloids.

Figure 1.4 shows two main variants of the parabolic cylindrical reflector telescope used as a meridian transit instrument. One is

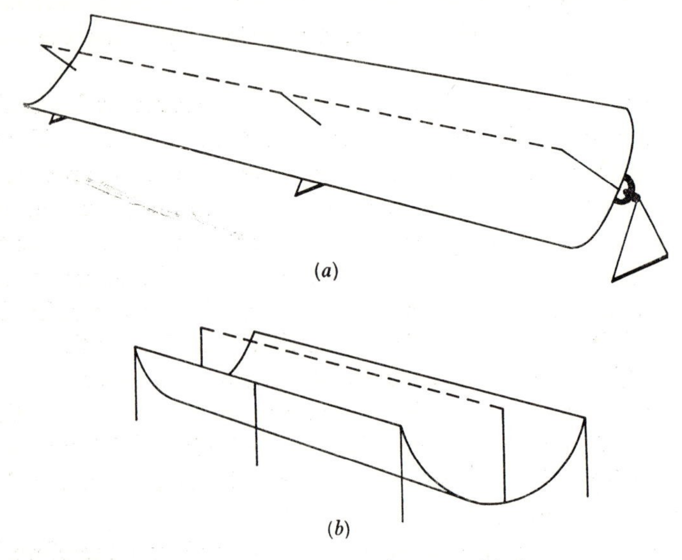

Figure 1.4. Parabolic cylinder reflecting telescopes. (*a*) An east–west antenna steerable in elevation. (*b*) A fixed north–south antenna with electrical steering of the antenna response.

lined up along an east–west axis and can be steered in altitude, but the electrical connections are fixed. The main axis of the other is north–south and the antenna response can be steered electrically by adding the outputs from the dipoles at the line focus in different phase combinations before sending the signals to the receiver. There are no moving parts and the reflector may, for example, be formed by a reflecting surface on a cylindrical excavation in the ground. This telescope has the disadvantage that it has a line focus rather than a point focus, and all the dipoles in the line must be changed if we want to work at a different wavelength. It is, on the other hand, much cheaper to construct than a fully steerable paraboloid of the same aperture, but, if we want a 'pencil beam' rather than a 'fan beam', then the aperture must have about the same dimensions east–west and north–south and much of the advantage in cost disappears.

Two other filled-aperture reflectors of the meridian transit type are shown in Figure 1.5. The Kraus antenna has a flat rotatable reflector that deflects energy arriving from a region situated somewhere along the meridian into a fixed paraboloid reflector and from there into a horn or other suitable feed antenna placed at the focus of the paraboloid. This antenna is analogous to the fixed horizontal telescope with a flat movable mirror sometimes used in optical astronomy.

The Pulkovo antenna has no movable plane mirror. The primary reflector consists of a large number of reflecting plates which can be individually adjusted to form part of the surface of a large paraboloid. The beam can be moved in the meridian plane by readjusting positions and tilts of all the individual reflecting plates so that they form part of another paraboloid.

A spherical reflecting surface is sometimes used in radio telescopes. Unlike the paraboloid, it has no main optical axis and can receive radio waves from a great range of angles without being moved (Figure 1.6*a*). The locus of the quasi-focus of the rays is a concentric spherical surface of half the radius of the reflector and the feed may be moved over this surface to absorb the energy flux from different parts of the sky. Alternatively, many fixed feeds with individual receivers may be placed on this surface and used simultaneously so as to allow an 'image' of the sky to be constructed. It is not very economical to use only the small part of the reflector from which the rays can be brought to a focus without excessive spherical aberration. A radial line-feed or a specially shaped secondary mirror as shown in Figure 1.6*b* makes it possible to collect the radiation from a much larger part of the reflecting surface.

Another fixed-reflector type of telescope should be mentioned: the fixed-plane array of dipoles with a plane reflector and with electrical steering of the beam. This type is used only at metre and decimetre wavelengths. Large apertures (say, 10^4 wavelengths)

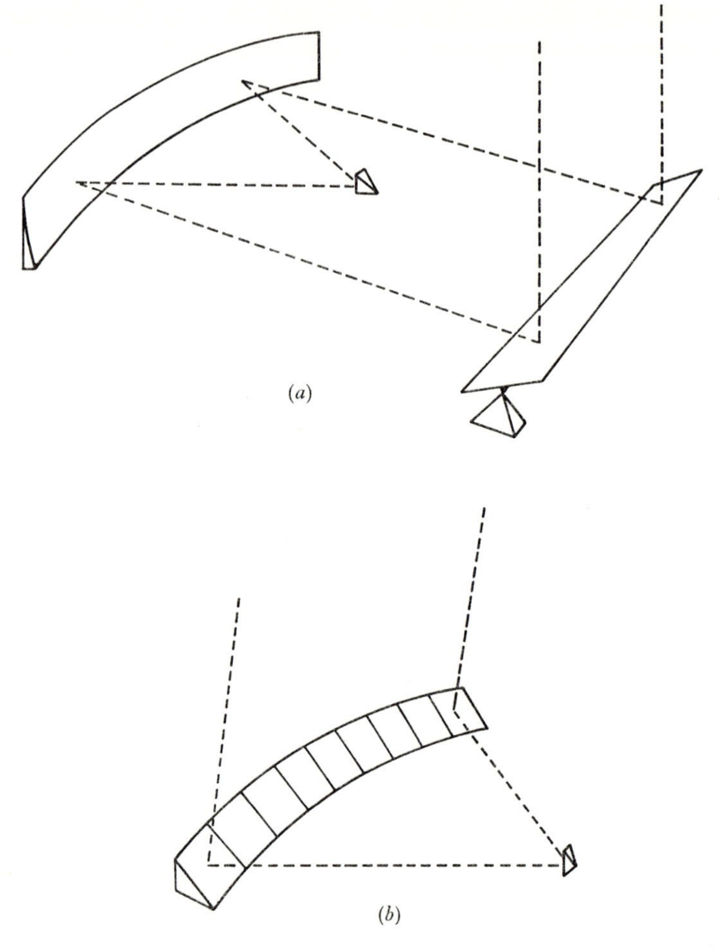

Figure 1.5. (*a*) A Kraus type reflecting antenna with a fixed parabolic surface and a rotatable flat surface. (*b*) The Pulkovo reflector is made up of movable panels and can be reformed for different angles of elevation of the antenna beam.

would require an impracticably large number of dipoles and connections ($\sim 10^8$).

1.4. Unfilled-aperture radiotelescopes

The reason for the term 'unfilled aperture' should be clear from the contents of this section: it implies that we can achieve an angular resolution corresponding to a large single (filled) aperture telescope while leaving out large portions of the aperture. The radio interferometer and the radio grating were used for this purpose quite early in the short history of radio astronomy. The interferometer has a special interest from a theoretical point of view because it constitutes the fundamental 'building block' of many later developed unfilled-aperture telescopes.

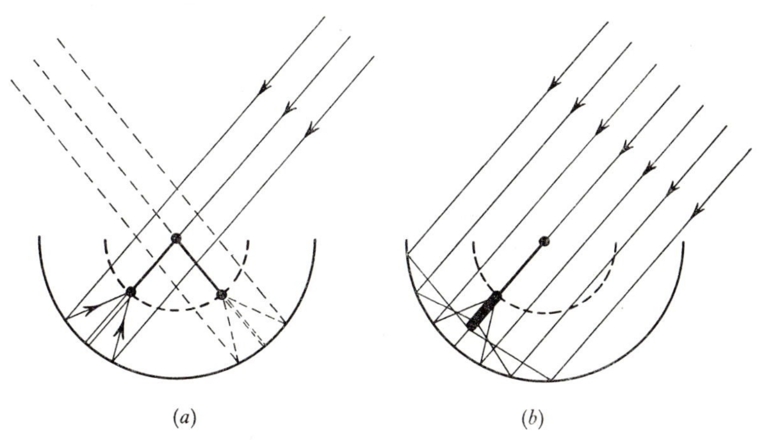

Figure 1.6. A fixed hemispherical surface may be used for reception in different directions by changing the position of the 'feed' antenna. (*a*) A simple feed may be used if only a limited part of the spherical reflector is used. (*b*) A more complicated feed may be used to correct the aberrations produced by the spherical reflector.

1.4.1. *The radio interferometer*

A (total power) radio interferometer consists of two antennas separated by a distance of L wavelengths and having the two outputs joined together at the input of a receiver (Figure 1.7). Waves from a direction $\cos\alpha = n/L$ will produce voltages which add at the receiver input. For directions $\cos\alpha = (n+\frac{1}{2})/L$ the voltages are in antiphase and cancel and no power reaches the receiver. The interference between the voltages from the two antennas causes a modulation of the normal power pattern as the Earth's rotation produces changes in α.

This interference pattern occurs only when the source is not large compared with the angular distance $1/L$ between adjacent 'fringes'. As the distance L between the two elements of an interferometer is increased, the interference pattern remains visible until the angular distance $1/L$ becomes less than the source size. Thus a variable-spacing interferometer can be used to determine the angular size of a source—it was, in fact, used for this purpose in optical astronomy many years before the start of radio astronomy.

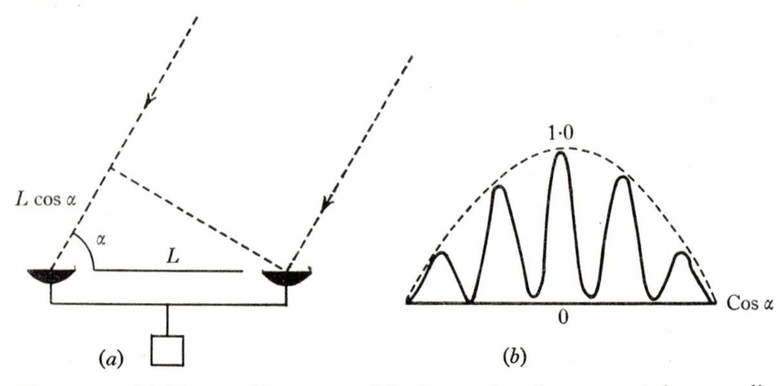

Figure 1.7. (*a*) Two aerials separated by L wavelengths, connected as a radio interferometer and receiving plane waves from a direction α. (*b*) The output from the interferometer as a function of the position of the source.

1.4.2. *The skeleton radiotelescope*

If radiation is falling on an aperture, we can imagine that the aperture is composed of a grid of small elements, each of which is collecting energy and transferring it to a common receiver where all the individual contributions are added (Figure 1.8). At the receiver the contribution from each element will produce, separately, an interference pattern with each of the other $N-1$ elements. The overall effect can be viewed as $N(N-1)/2$ superposed interference patterns. The spacing L of the elementary interferometers ranges from very small distances to the maximum distance available in the aperture and all orientations of the axis are represented.

In the square aperture of Figure 1.8*b* the $N-1$ interference patterns of X with all others plus the $N-2$ of Y with all others cover all the distances and directions that were found in the $N(N-1)/2$ patterns of Figure 1.8*a*. These $2N-3$ patterns must, then, contain the same information as the $N(N-1)/2$ patterns. From this it appears that there is, in a certain sense, a great amount of redundancy in the information gathered by a filled-aperture antenna and that it should be possible to devise an aperture in which all the essential spacings and axis orientations for the necessary interference patterns are present, but for which the total area is much reduced. This, in fact, is the basis of the skeleton antennas which have been developed by radio astronomers. In Figure 1.9*a* we have two strips forming an inverted T. The vertical strip *A* is cut up into $\sqrt{N}$ sections and the horizontal strip *B* into twice that

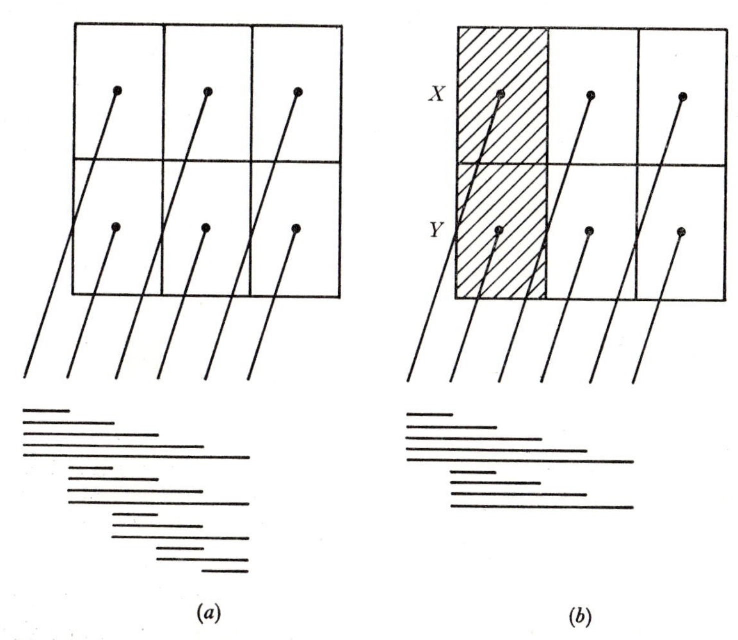

Figure 1.8. A rectangular antenna considered as an assembly of smaller elements. (*a*) The spacings between elements taken two at a time are shown in the lower part of the diagram. (*b*) The spacings between element *X* and all the others plus those of element *Y* and all the others.

number. There are $2N$ different interferometer pairs that can be made from each element of *A* with each element of *B*. These $2N$ interferometer pairs represent all the spacings and distances found in the $N(N-1)/2$ interferometers of the filled aperture *C* of Figure 1.9*a*. In a similar way it can be seen that the regular matrix *B* of small elements, shown in Figure 1.9*b*, when combined with one large element *A*, which has the dimensions of the matrix element spacing, produces an interferometric system in which are represented all spacings and directions that would be found in a square filled aperture having dimensions equal to $B/2$.

1.4.3. *Aperture synthesis*

The response of an interferometer to a non-varying radio source does not change from day to day and it follows that the information about the brightness distribution over the source can be derived from the elementary interferometer patterns *taken one at a time*—i.e. it is not necessary to have all the elements of Figure 1.9*a* or 1.9*b* present simultaneously. We could move one small element of A along the arm A and one element of B along the arm B, obtaining the interference patterns in each position of A with every position

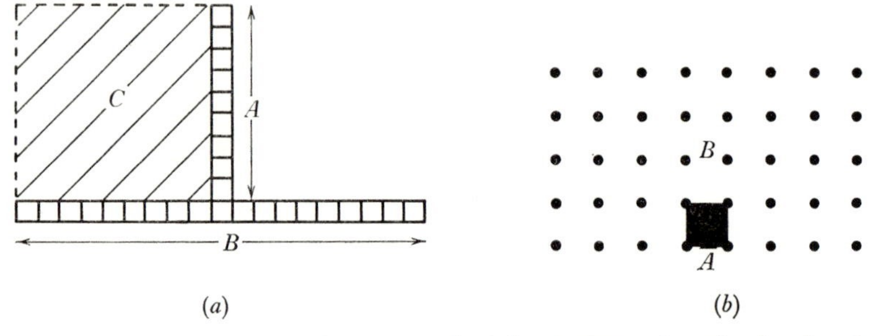

Figure 1.9. (*a*) The spacings between each of the elements of a strip A and each of the elements of strip B represent all the spacings found in the rectangle C. (*b*) The same range of spacings found in the arrangement (*a*) is found in the spacings between A and each of the elements of B.

of B. There are $2N$ separate patterns, which contain all the information that was present in the filled aperture of Figure 1.8 or the skeleton of Figure 1.9*a*. Hence, if we are prepared to spend $2N$ times longer in getting our observations, we can, with only two small elements, make the same observations (with the same sensitivity for a point source) as with an area $2\sqrt{N}$ times larger (Figure 1.9*a*). There is, however, another factor which makes the *aperture-synthesis* antenna more attractive. A small element with an area of a square wavelength collects energy from a solid angle of the sky of about $1/a$ steradians, since the beamwidth is inversely proportional to the linear dimensions. The $2N$ observations, then, contain information about the brightness distribution over the whole region of $1/a$ steradians. With the filled aperture of Figure 1.8 or the skeleton antenna of Figure 1.9 we usually have combined the outputs of the elementary parts of the system into a *single* output. This makes it an N times larger aperture, which collects energy from an N times smaller region of sky, i.e. $1/(Na)$ steradians. If we want to make a study of the region $1/a$ we must make N separate observations—i.e. we must scan the region. The time required to do this reduces the unfavourable ratio of the time of observation with aperture synthesis compared with the skeleton antenna of Figure 1.9*a* from $2N$ to 2. It is possible, of course, to improve the skeleton antenna by con-

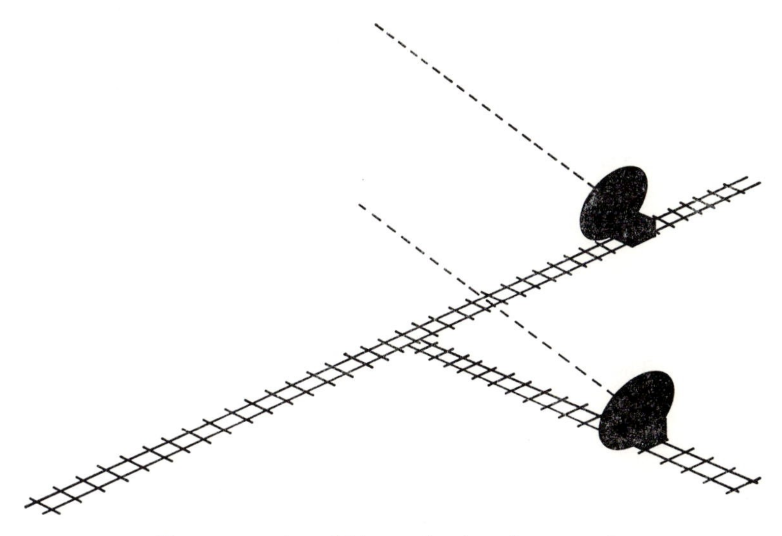

Figure 1.10. A variable-spacing interferometer for two-dimensional aperture synthesis.

necting all elements separately to a receiving system and by extracting all elementary interference patterns separately through $2N$ separate channels. By this means the ratio of times of observation of the two systems is again increased to $2N$ because the systems are the same, except that the observations are made simultaneously with one system and serially with the other (Figure 1.10).

There is a second way in which an aperture may be synthesized. If a line-aperture is rotated, a two-dimensional aperture may be generated. It is often convenient to use the Earth's rotation (Figure 1.11) to synthesize the aperture from a line-aperture. Antennas which make use of this principle will be described in Chapter 7.

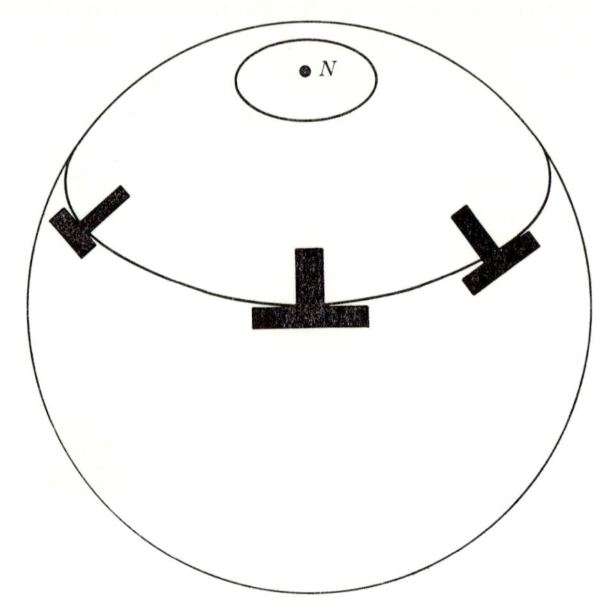

Figure 1.11. The apparent rotation of an area of the Earth's surface during the day as seen from a fixed point in space.

This apparent rotation of a line on the Earth forms the basis of one method of aperture synthesis.

1.4.4. *Antennas with an incomplete range of spacings and directions*

A line-aperture is incomplete inasmuch as it has a full range of spacings in only one direction. As we have described, the range of spacings and directions may be completed by rotating the line-aperture. The line-aperture itself can be of the skeleton form or may be synthesized by a variable-spacing interferometer. The most commonly used skeleton line antenna is the compound grating antenna, which consists of a grating *A* (Figure 1.12) combined with a large single antenna *B* having dimensions comparable with the gaps in the grating. We shall describe such line antennas in Chapter 6.

It should be pointed out that it is not always necessary to have a continuous range of spacings and directions in an aperture. When strong isolated sources are being studied we can tolerate gaps in the aperture—that is, we can 'sample' spacings and directions. The effect of the gaps or sampling procedure is to form multiple responses in the sky, but these need produce no ambiguities if a single strong source is under observation.

With very high resolution observations it is economical to use a minimum sample of the aperture. The earliest antenna in which this was done was the simple grating antenna (*A* of Figure 1.12) used for solar observations. Later the two-dimensional form of this, the grating cross, was used for the same purpose.

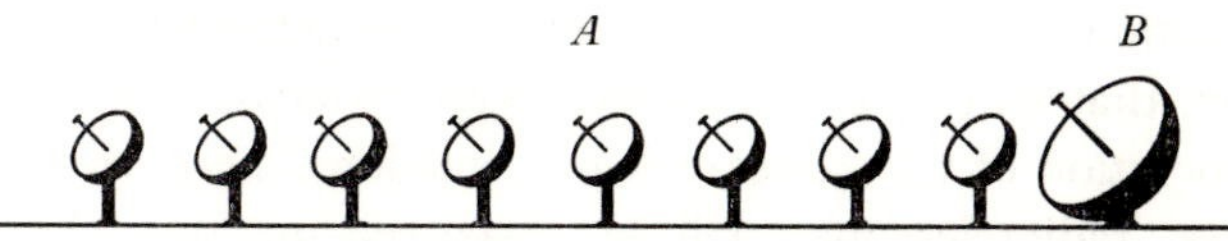

Figure 1.12. A 'skeleton' line antenna (compound grating). Seen from the direction in which the antennas are pointing, it contains the spacings found in a continuous line antenna of the same length.

It consists of a grating antenna *A* and a larger antenna *B* which has a diameter greater than the spacing between grating elements.

1.4.5. *Independence of sensitivity and resolution*

Unfilled-aperture antennas have the important property that their angular resolution and effective area can be chosen independently. This means considerable economy in materials without any reduction in the efficiency of the telescope. It is not possible with filled-aperture telescopes, for which resolving power and effective area are not independent.

CHAPTER 2

SOME THEORY

The theory of radiotelescopes rests, of course, on the foundations of electromagnetic theory and on developments in both optical diffraction theory and radio-antenna theory. Only those parts of the theory that are of direct relevance to an understanding of the mode of operation of telescopes will be considered here. Simple 'total-power' radiotelescopes will be discussed in this chapter; the correlation type telescopes will be treated later.

2.1. Effective area and antenna polarization

A total-power radiotelescope may consist of a single antenna or it may be an array of several antennas connected together electrically to form a single unit. The power per unit frequency interval p(W Hz^{-1}) which is available at the antenna output terminals will clearly be proportional to the energy flux per unit frequency interval, i.e. the *flux density* S_m(Wm^{-2} Hz^{-1}) of the source that is being measured (the subscript m indicates that the polarization of the received radiation should be 'matched' to that of the antenna. The concept of matched polarization will be explained later in this section). Thus

$$p = A \, . \, S_m \quad (\text{WHz}^{-1}) \tag{2.1}$$

The proportionality constant A is the *effective area* of the radiotelescope and expresses how effective the telescope is at absorbing radiation of this frequency from any particular source direction and in making this power available at the output terminals. It has the dimension of area (m^2) and the equation shows that the available power per unit frequency interval is equal to that passing through any area A at right angles to the direction of the source. It does *not* imply, however, that the power p is necessarily derived from any particular physical area A, although in special circumstances A may be approximately equal to the physical aperture of the antenna.

The actual power in watts measured by the receiver will be proportional to the range of frequencies or the frequency *bandwidth*

$\Delta\nu$ (Hz) over which the measurements are made. The bandwidth is determined either by the radiation itself, which may, for instance, be confined to a narrow spectral line, or, more usually, by the antenna-receiver system which accepts signals only within a limited frequency range. The receiver is usually designed to have a smaller bandwidth than the radiation itself, so that information about the shape of the source spectrum is not lost through averaging over too great a frequency interval. The bandwidth, therefore, is usually an instrumental parameter and the available power is interpreted as that within the accepted frequency band independent of which part of the system does, in fact, determine the bandwidth. In the simple case of a telescope with a narrow band $\Delta\nu$ over which all parameters, including the flux density of the source, may be regarded as constant, we can write the available power W:

$$W = p \, . \, \Delta\nu = A \, . \, S_m \, . \, \Delta\nu \quad \text{(Watt)}. \tag{2.2}$$

Radio-frequency radiation, like light, consists of transverse electromagnetic waves, and the field vector is confined to a plane parallel to the wavefront. The wave is elliptically polarized if the end of the field vector consistently traces out an ellipse with constant eccentricity and orientation. Linear and circular polarization are special cases of elliptical polarization. If, on the other hand, the path traced out by the field vector is continually changing in a random fashion, we are dealing with a randomly polarized, or 'unpolarized', wave.

A simple dipole antenna will react only to the vector component of the electric field that is parallel to the dipole itself: the dipole is a linearly polarized antenna. The orthogonal vector component of the field, which represents half the total energy in a randomly polarized wave, cannot give rise to any currents and does not, therefore, interact with the dipole. In order to absorb all the power in a randomly polarized wave we need two orthogonal dipoles with independent outputs, each interacting with one vector component of the field. The wave field has two degrees of freedom, since it is defined by a vector in a plane, and two independent output voltages will, in general, be needed to convey all the information contained in a single wave. Consequently, it is not possible to combine the two outputs into a single output which delivers all the available power from the two dipoles. The combination will accept signal components which are in phase and add at the junction, but those which are in antiphase will cancel and the power they represent will be reflected back to the dipoles and reradiated into space. The junction, in fact, defines a particular phase and amplitude relation between the signals from the two dipoles, i.e. a particular polarization of an incoming wave which will be accepted and a complementary one which will be rejected. The accepted or *matched polarization* may be linear, circular or elliptical, depending on how the junction has been constructed. Apart from being differently polarized the combination of the two dipoles acts as a single dipole in accepting a matched polarized component of the radiation and rejecting the complementary polarized component.

The antenna polarization properties illustrated by the dipoles can be extended to cover all types of antennas. We state here without proof some more general theorems which are relevant to antenna polarization properties:

(1) Any radio wave can be resolved into two polarized components. The polarization of one component can be chosen at will and the complementary polarization will then be characterized by a similar polarization ellipse with the major axis at right angles to that of the first component and with the electric vector rotating in the opposite sense.

(2) The two polarized components carry, on the average, equal amounts of power in a randomly polarized wave. They will, in general, carry different powers if the radiation contains some non-randomly polarized components.

(3) Any antenna or antenna system with a single output (i.e. any 'total-power' radiotelescope) is sensitive to radiation of one particular polarization which is matched to the antenna. The available power at the output is proportional to the flux density of the matched polarized component of the incoming radiation and is independent of the flux density in the complementary polarized component.

The total flux density S of a radio source must therefore always be regarded as the sum of one matched polarized component S_m which contributes to the available power at the antenna output and one complementary polarized component S_c which does not. It is clear that the definition of antenna effective area (2.1) must be expressed in terms of the former component only. If the radiation is known to be randomly polarized we can express S_m in terms of S since

$$S_m = S_c = \tfrac{1}{2}S \quad \text{(randomly polarized waves).} \tag{2.3}$$

Much of the natural radiation from space has some (non-randomly) polarized component and (2.3) will not, in general, be valid. The correct procedure for arriving at the flux density S of a source is to measure separately the flux density at two complementary polarizations.

If an antenna delivers to the receiver all the matched polarized energy falling on its surface, then its effective area will clearly be equal to its geometrical area. In practice the effective area in the direction of maximum response $A_{\max}$ sometimes approaches the geometrical area in magnitude, but in other directions it will be smaller. We shall write the effective area as $A(l, m)$ to indicate that it is a function of the direction of arrival of the waves (l, m). The effective area, normalized to unity in the direction of maximum response, is the *power pattern* ('beam') of the antenna.

$$P(l, m) = A(l, m)/A_{\max}. \tag{2.4}$$

At this point it is necessary to introduce the coordinate systems that will be used. The antenna will be described in terms of the rectangular cartesian coordinates (x, y, z) and usually the x, y plane will be chosen so that it is directly related to the antenna aperture plane. The unit of length will be one wavelength at the centre frequency of the band accepted by the receiver.

For convenience we shall specify a direction by its directional cosines (l, m, n) with respect to the (x, y, z) axes. Alternatively we may define the direction in terms of the rectangular coordinates (l, m, n) of a point on a celestial sphere of unit radius (Figure 2.1). Since l, m, n are related by $l^2 + m^2 + n^2 = 1$ it is sufficient to use only the coordinates (l, m), the third coordinate being understood. l and m are therefore rectangular coordinates by which we can describe the projection of the celestial sphere onto the l, m plane. Each point (l, m) actually corresponds to two directions, one in each hemisphere, but generally this will cause no confusion. A small solid angle $d\Omega$ in the direction (l, m) projects on the l, m plane so that

$$d\Omega = \frac{dl\,dm}{\cos\gamma} = \frac{dl\,dm}{(1 - l^2 - m^2)^{\frac{1}{2}}}. \tag{2.5}$$

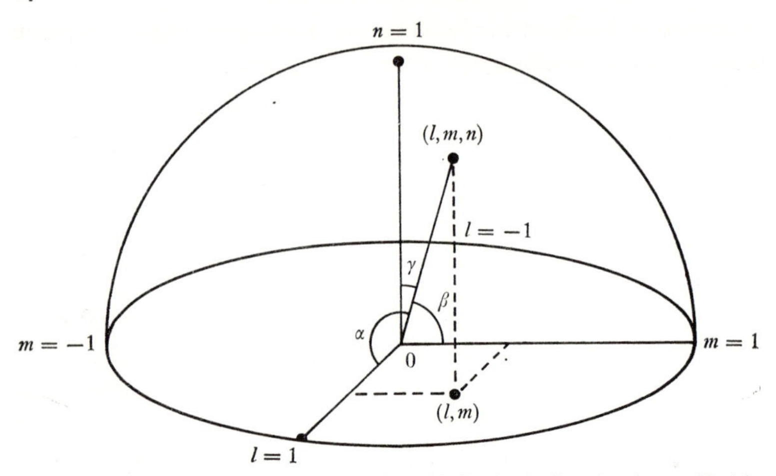

Figure 2.1. The coordinate system used in this book. A direction is specified by the directional cosines l, m, n, i.e. $\cos\alpha$, $\cos\beta$, $\cos\gamma$, where only two of the three quantities can be chosen independently. Any direction in the celestial hemisphere may be represented by a point on a unit sphere or by its projection (l, m) on the l, m-plane.

The relations between the (l, m, n) coordinates and other coordinate systems are shown in Appendices 1 and 4.

2.2 Radio brightness, antenna temperature, brightness temperature and some simple associated antenna relations

The effective area $A(l, m)$ was defined in terms of power flux from a 'point' source in the direction (l, m). However, in practice there will be some radiation arriving from all directions. The radiation from different parts of space is uncorrelated and it follows that the mean square voltage $\langle V^2 \rangle$ at the output terminals is the sum of the individual mean square voltages due to the radiation from different directions—i.e. available powers at the antenna output add linearly. We shall derive p, the available power per unit bandwidth, when the telescope is used to measure an extended source characterized by the *radio brightness* distribution $B(l, m)$. This describes the flux density per unit solid angle of the sky as a function of the direction (l, m):

$$B(l, m) = \Delta S/\Delta\Omega \quad (\mathrm{W\,m^{-2}\,Hz^{-1}\,sterad^{-1}}). \tag{2.6}$$

The matched polarized component of the radiation has the radio brightness

$$B_m(l, m) = \Delta S_m/\Delta\Omega. \tag{2.7}$$

In the special case of a randomly polarized source we can write in analogy with (2.3)

$$B_m = B_c = \tfrac{1}{2}B \quad \text{(random polarization)}. \tag{2.8}$$

From (2.1) and (2.7) we can derive the contribution Δp W Hz^{-1} due to the radiation from the solid angle $\Delta\Omega$ in the direction (l, m):

$$\begin{aligned}\Delta p &= A(l,m) \,.\, \Delta S_m \\ &= B_m(l,m) \,.\, A(l,m) \,.\, \Delta\Omega \quad (\text{W Hz}^{-1}).\end{aligned} \tag{2.9}$$

Since the power contributions from all directions add linearly, the total power per unit bandwidth due to radiation received from the sky is obtained by integrating (2.9) over the whole sky.

$$p = \int_{4\pi} B_m(l,m) \,.\, A(l,m) \, d\Omega \quad (\text{W Hz}^{-1}). \tag{2.10}$$

p is often equated to the thermal noise power available from a resistor at a temperature T_a °K, with the resistor replacing the antenna at the input of the receiver. The temperature T_a which corresponds to a power per unit bandwidth p is given by Nyquist's formula

$$p = kT_a \quad (\text{W Hz}^{-1}), \tag{2.11}$$

where $k \approx 1{\cdot}38 \,.\, 10^{-23}$ W Hz^{-1} degree^{-1} is Boltzmann's constant. The temperature T_a of this fictitious resistor is called the *antenna temperature*. The unit represented by an antenna temperature $T_a = 1$ °K, or $1{\cdot}38 \,.\, 10^{-23}$ W Hz^{-1}, is much more convenient in radio astronomy than the much larger unit for p, the W Hz^{-1}. It should always be kept in mind, however, that the temperature T_a simply indicates a power level and need have no relation to any real temperature.

In an analogous way radio brightness can also be expressed in terms of a temperature. In this instance we equate the radio brightness (flux density per unit solid angle) of the source in some direction (l, m) with that of a black-body radiator at a temperature T_B °K. The temperature T_B of this fictitious black-body radiator is the *brightness temperature* of the source in this particular direction. The relation between a radio brightness B and the corresponding brightness temperature T_B is given by Rayleigh's classical approximation to the radiation law

$$B \approx 2kT_B/\lambda^2 \quad (\mathrm{W\,m^{-2}\,Hz^{-1}\,sterad^{-1}}), \tag{2.12}$$
$$(\text{for } h\nu \ll kT_B).$$

The thermal radiation from a black body is randomly polarized and it follows that the matched polarization brightness $B_m = \frac{1}{2}B$ (2.8). T_B, like B_m, will in general be a function of the direction (l, m) and we get

$$B_m(l,m) = kT_B(l,m)/\lambda^2 \quad (\mathrm{W\,m^{-2}\,Hz^{-1}\,sterad^{-1}}). \tag{2.13}$$

It should be stressed that (2.13) defines the radio brightness in terms of a black-body temperature and this temperature need have no relation to any real temperature. If this is kept in mind, then no difficulties arise when the 'temperature' is different for antennas of different polarization. If we substitute T_a and $T_B(l, m)$ in (2.10) for p and $B(l, m)$ we have

$$T_a = \lambda^{-2} \int_{4\pi} T_B(l,m)\, A(l,m)\, d\Omega, \tag{2.14}$$

which shows that the antenna temperature equals the all-sky integral of the brightness temperature weighted by $A\lambda^{-2}$, i.e. by the antenna effective area (expressed in square wavelengths). An important relation can be derived from this. Let the antenna be terminated by a matched resistor and the whole system placed in a black box at a real temperature T °K. The system is in equilibrium and as much energy kT flows into the resistor via the antenna as flows out of it. Thus the antenna temperature $T_a = T$. But by definition $T_B(l, m)$ is everywhere also equal to T, i.e.

$$T_a = T_B(l,m) = T \tag{2.15}$$

and substituting these values in (2.14) we find

$$\int_{4\pi} A(l,m)\, d\Omega = \lambda^2, \tag{2.16}$$

or that the *all-sky integral of the effective area is one square wavelength*. An important consequence of this is that the effective area is determined in theory from the shape of the reception pattern alone. Let this be represented by the power pattern (2.4). Equation

(2.16) then leads to
$$A_{\max} = \lambda^2 \Big/ \iint_{4\pi} P(l,m)\, d\Omega. \tag{2.17}$$

The effective area is then given by
$$A(l,m) = A_{\max} \cdot P(l,m). \tag{2.18}$$

The equations derived so far are valid only for ideal antennas with no internal ohmic losses. If a fraction $(1-\eta_R)$ of the energy that in an ideal case would be fed to the receiver is absorbed by the antenna structure itself, then the calculated effective area must be multiplied by the so-called *radiation efficiency* factor η_R. This and other effects of internal losses will be discussed later, in the chapter on Sensitivity.

2.3. Field pattern and grading

In the previous sections we have discussed the available power at the antenna output, the radiation from the sky and how these are related through the effective area of the antenna. We shall now discuss how this effective area is determined from the current distribution over the antenna. In order to do so we need to introduce two new concepts, the field pattern $F(l,m)$ and the grading $g(x,y)$.

The *field pattern* is a function, in general complex, which describes the amplitude and phase of the voltage at the terminals of the antenna as a function of the direction of arrival of waves from a distant source of constant strength but variable direction. The phase term in the function is the phase relative to that at the terminals of an isotropic (imaginary) antenna at the origin of the (x,y,z) coordinate system on the ground. The field pattern is normalized to unit amplitude $[|F(l,m)| = 1]$ in the direction of maximum response, i.e. where $A(l,m) = A_{\max}$. Since power is proportional to voltage squared we have (2.18)

$$\begin{aligned} A(l,m) &= A_{\max} \cdot P(l,m) \\ &= A_{\max} \cdot |F(l,m)|^2 \quad (\text{m}^2). \end{aligned} \tag{2.19}$$

The fundamental approach to the problem of determining the field pattern of a voltage-fed antenna is by means of Maxwell's equations with boundary conditions fixed by the metallic or dielectric structure of the antenna. This approach is possible only in the simplest cases. In practice, therefore, approximations must be introduced. When applying such approximate methods it is advantageous to treat the antenna as transmitting instead of receiving. The reciprocity theorem states that the antenna characteristics such as the field pattern are identical when transmitting and when receiving.(201) The field pattern can therefore be defined in an equivalent way by the transmitted wave in the far field when normalized to unity in the maximum direction. The phase of the field pattern, represented by the argument of the complex function $F(l, m)$, is defined as the difference in phase between the transmitted wave and an imaginary spherical reference wave derived from the same generator but transmitted from the x, y, z origin. The far field can be calculated by integrating the contributions from all current elements in the radiating parts of the antenna.

If we think of an antenna as being completely enclosed in a surface over which the fields (or currents) are known, then we can calculate the external fields.(202) The required surface may for simplicity, be taken as an infinite plane immediately in front of the antenna if this radiates into one hemisphere only. We can often make the further approximation that the currents in this *aperture plane* are limited to a region corresponding to the physical size of the antenna, the *antenna aperture*. The 'aperture' has some physical significance for many common antennas such as horns or paraboloids, at least when it is very large compared with the wavelength. The aperture plane can be chosen as the x, y-plane; all currents are then at $z = 0$ (Figure 2.2).

The field $\delta E(l, m) \,.\, \exp(j\omega t)$ in the transmitted wave measured at some large distance d from the origin and produced by the current in a small aperture element $\delta x\, \delta y$ can be written

$$\delta E(l, m) = \text{const.}\,.\, I_0\, g(x, y)\, \delta x\, \delta y\,.\, F_e(l, m) \exp\{j2\pi(xl + ym)\}. \tag{2.20}$$

The frequency factor $\exp(j\omega t)$ has been omitted from both sides of the equation. The quantities appearing on the right-hand side are as follows:

The *grading* $g(x, y)$ describes the current distribution over the aperture. It is in general a complex function expressing the amplitude and the phase of the current and is usually normalized to unit amplitude at the point of maximum current density. The expression $I_0 g(x, y)\,\delta x\,\delta y$ (amp m) is the current moment carried by the aperture element $\delta x\,\delta y$. $F_e(l, m)$ is the *field pattern of an aperture element* situated at the x, y, z-origin. If, for instance, this is linearly polarized parallel to the x-axis (equivalent to a short or Hertz dipole) then F_e will equal the cosine of the angle α with the x-axis

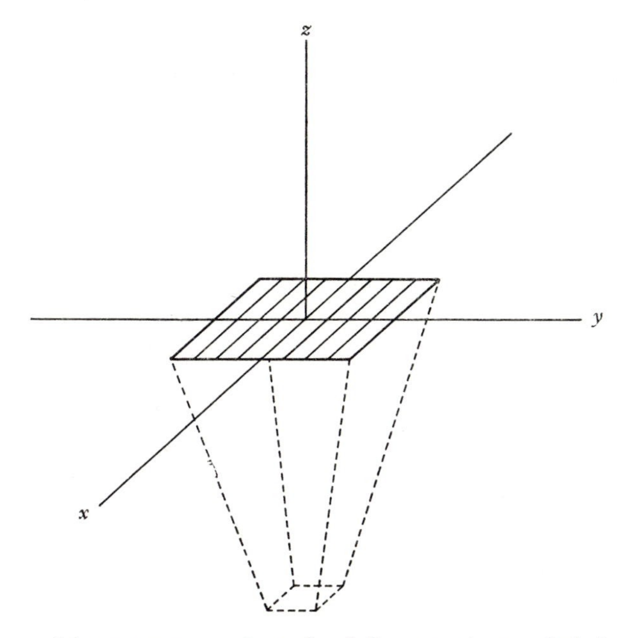

Figure 2.2. A horn antenna can be replaced, for approximate calculations, by a current distribution over a limited area in the x, y-plane. This area is called the aperture of the horn.

or, in the l, m coordinates, $F_e(l, m) = \sqrt{(1 - l^2)}$. The field pattern of an element at (x, y, z) will have the same amplitude in all directions, but the phase, which is defined relative to a spherical reference wave from the origin, will differ by $2\pi(xl + ym + zn)$ radians as shown in Figure 2.3. Equation (2.20) is now seen to state simply that the far field due to the aperture element at (x, y, 0) is proportional to the product of its current moment $I_0 g(x, y)\,\delta x\,\delta y$ and its field pattern $F_e(l, m) \exp\{j\,2\pi(xl + ym)\}$.

By integrating (2.20) over the whole antenna aperture we get the total far field $E(l, m)$ at the distance d from the origin. The (transmission) field pattern was defined as the far field, normalized to an amplitude of unity in the maximum direction. Hence

$$F(l, m) \propto E(l, m)$$

and, collecting the various constants together, we get:

$$F(l, m) = \text{const.}\, F_e(l, m) \iint_{-\infty}^{+\infty} g(x, y) \exp\{j2\pi(xl+ym)\}\, dx\, dy, \tag{2.21}$$

where the constant has the value which makes $|F(l, m)|_{\max} = 1$. The integration over the aperture has been formally replaced by an

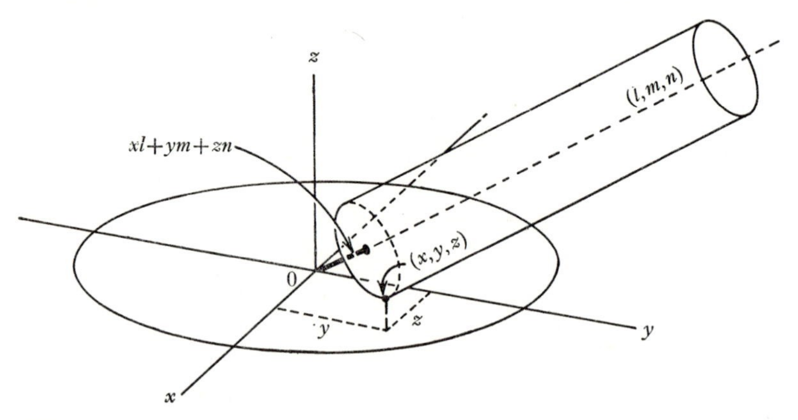

Figure 2.3. The path differences between a wave from the origin (0, 0, 0) and a wave originating from a point (x, y, z) at a very distant point in the direction l, m, n.

integration over the whole x, y-plane; this does not change the value of the integral if we make the obvious definition $g(x, y) = 0$ for the region outside the aperture.

The extension of (2.21) to three dimensions is straightforward: the double integral is replaced by a triple integral, the grading is a function of all three coordinates x, y, z and the exponential contains the factor $(xl+ym+zn)$ where, however, it should be remembered that n is not an independent variable but related to l and m by the equation $l^2+m^2+n^2 = 1$. The three-dimensional formula is needed for accurate calculations on antennas such as the parabolic reflector for which the 'aperture plane' treatment is a rather coarse approximation.

In special cases the currents may be regarded as being confined to a straight line and we can use the simple one-dimensional form of (2.21).

$$F(l,m) = \text{const. } F_e(l,m)\int_{-\infty}^{+\infty} g(x)\exp\{j2\pi xl\}\,dx. \qquad (2.22)$$

The integrals in (2.21) and (2.22) will be recognized as a two-dimensional and a one-dimensional Fourier integral respectively. The many remarkable properties of such integrals (see Appendix 2) can be used to facilitate antenna calculations and, in fact, Fourier theory plays an important part in all such work[203–5].

$f(l)$, the one-dimensional Fourier transform of $g(x)$ is defined by the equation

$$f(l) = \int_{-\infty}^{+\infty} g(x)\exp\{j2\pi xl\}\,dx. \qquad (2.23)$$

If $f(l)$ is given by (2.23), then $g(x)$ can itself be written as the inverse Fourier transform of $f(l)$

$$g(x) = \int_{-\infty}^{+\infty} f(l)\exp\{-j2\pi xl\}\,dl. \qquad (2.24)$$

The two functions are related in a nearly symmetrical way and are said to constitute a *Fourier pair* for which we shall use the symbol $\mathscr{F}$. Thus

$$f(l)\,\mathscr{F}\,g(x) \qquad (2.25)$$

is equivalent to (2.23) and implies (2.24). If we want to stress the relation (2.24) rather than (2.23) we write the same Fourier pair as

$$g(x)\,\mathscr{F}\,f(l). \qquad (2.26)$$

Two-dimensional Fourier pairs obey equations directly analogous to (2.23) and (2.24).

2.4. Computing antenna patterns: the Fourier transform

We have discussed the equations that are needed in order to calculate the field pattern, the power pattern and the effective area of an antenna whose grading (i.e. current distribution when transmitting) is known. The general procedure for a two-dimensional aperture is as follows:

We first compute $f(l,m)$, the two-dimensional *Fourier transform* of the grading $g(x,y)$:

$$f(l,m) = \iint_{-\infty}^{+\infty} g(x,y)\exp\{j2\pi(xl+ym)\}\,dx\,dy \qquad (2.27)$$

and derive the *field pattern* from (2.21):

$$F(l,m) = \text{const.}\,F_e(l,m)\,.\,f(l,m), \tag{2.28}$$

where the constant should have the value which normalizes the field pattern (modulus) to unity in the maximum direction. The *power pattern* is the square of the field pattern modulus (2.19) and the *effective area* can finally be obtained from (2.17) and (2.18):

$$P(l,m) = |F(l,m)|^2, \tag{2.29}$$

$$A(l,m) = A_{\text{max}} P(l,m), \tag{2.30}$$

where
$$A_{\text{max}} = \eta_R \lambda^2 \Big/ \iint_{4\pi} P(l,m)\,d\Omega. \tag{2.31}$$

η_R is the 'efficiency factor' mentioned in connection with (2.17) and (2.18), which were derived for the ideal case of an antenna with no internal losses. The effective area as defined in section 2.1 (eqn 2.1) will clearly be smaller than this ideal value by the factor η_R if a fraction $(1-\eta_R)$ of the power absorbed by the antenna is dissipated in the antenna itself before reaching the output terminals.

In simple cases when we can regard the currents as being confined to a line along the x-axis and (2.22) is applicable, we can replace the two-dimensional transform (2.27) by the simpler one-dimensional Fourier transform

$$f(l) = \int_{-\infty}^{+\infty} g(x)\,.\,\exp\{j2\pi xl\}\,dx \tag{2.23}$$

and the field pattern is derived from (2.22):

$$F(l,m) = \text{const}\,F_e(l,m)\,.\,f(l). \tag{2.32}$$

2.4.1. *A uniformly graded line antenna*

An antenna L wavelengths long lies along the x-axis and the current, when transmitting, is uniform in amplitude and phase along the line. The grading is (see Figure 2.4*a*)

$$g(x)\begin{cases} = 1\cdot 0 & \text{for} \quad |x| \leqslant \tfrac{1}{2}L, \\ = 0 & \text{elsewhere} \end{cases} \tag{2.33}$$

and we can use the simple one-dimensional formulae. The Fourier transform of $g(x)$ is (see Figure 2.4*a*)

$$f(l) = \int_{-\frac{1}{2}L}^{\frac{1}{2}L} 1\cdot 0 \exp\{j2\pi xl\}\,dx = L\,.\,\frac{\sin(\pi Ll)}{\pi Ll}. \tag{2.34}$$

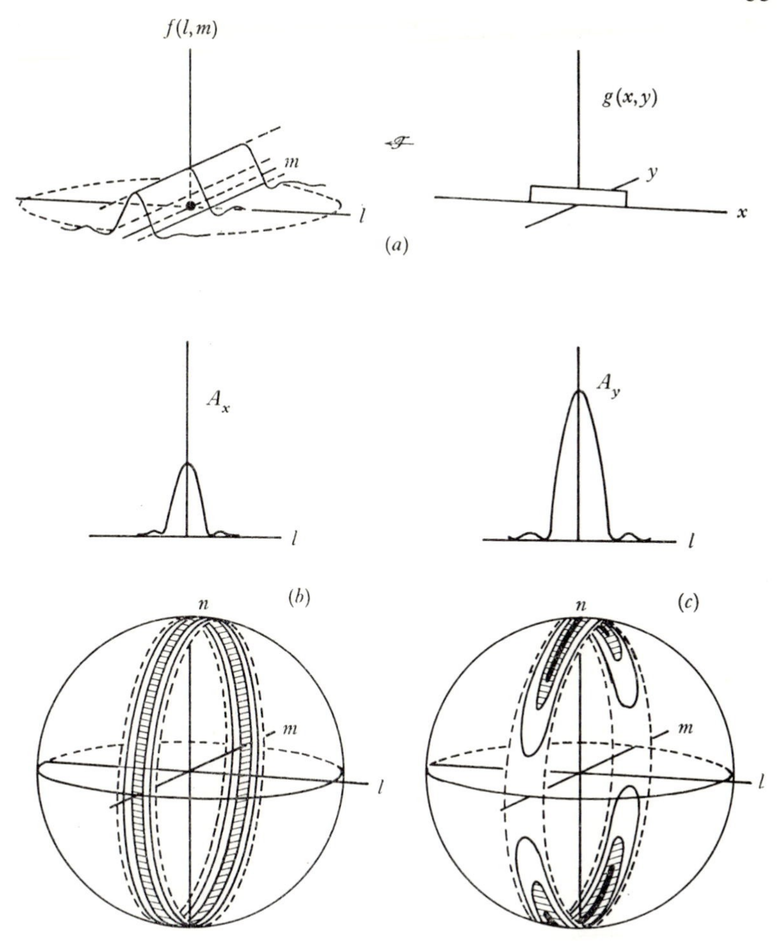

Figure 2.4. (*a*) The grading $g(x, y)$ of a line antenna and its Fourier transform $f(l, m)$. (*b*) The effective area $A_x(l, m)$ of the antenna when the current (polarization) is in the line of the antenna. (*c*) The effective area $A_y(l, m)$ of the antenna when the polarization is perpendicular to the line of the antenna.

The field pattern is (2.32)

$$F(l,m) = F_e(l,m) \,.\, \frac{\sin(\pi L l)}{\pi L l} \tag{2.35}$$

which is normalized to unity in the maximum direction (in this case the zenith). The pattern extends over a large angle of sky in the *m* coordinate, where it is determined only by the slowly varying element field pattern. If the antenna is linearly polarized parallel to the line, then (see section 2.3)

$$F_{ex}(l,m) = \sqrt{(1-l^2)} \quad \text{(parallel polarization)} \tag{2.36}$$

and the field pattern of the parallel polarized line antenna is:

$$F_x(l,m) = \sqrt{(1-l^2)}\frac{\sin(\pi Ll)}{\pi Ll}$$
$$\approx \frac{\sin(\pi Ll)}{\pi Ll} \quad (\text{for } L \gg 1). \tag{2.37}$$

If the antenna is linearly polarized at right angles to the line we have instead

$$F_{ey}(l,m) = \sqrt{(1-m^2)} \quad \text{(perpendicular polarization)} \tag{2.38}$$

and
$$F_y(l,m) = \sqrt{(1-m^2)}\frac{\sin(\pi Ll)}{\pi Ll}. \tag{2.39}$$

Solving (2.29 to 2.31) for the effective area we get for the parallel and perpendicular polarization respectively:

$$A_x(l,m) = \eta_R \frac{\lambda^2 L}{2\pi}\left\{\frac{\sin(\pi Ll)}{\pi Ll}\right\}^2 \quad (\text{for } L \gg 1), \tag{2.40}$$

$$A_y(l,m) = \eta_R \frac{\lambda^2 L}{\pi}(1-m^2)\left\{\frac{\sin(\pi Ll)}{\pi Ll}\right\}^2. \tag{2.41}$$

Figures 2.4*b* and 2.4*c* illustrate the effective areas as given by these equations. Perpendicular polarization gives twice as large effective area in the maximum direction, but the whole pattern is restricted to a smaller region of the sky, making the all-sky integrals the same in the two cases.

2.4.2. *A uniformly graded large rectangular aperture*

Let the sides of the aperture be L_x and L_y wavelengths; its two-dimensional grading is then

$$g(x,y)\begin{cases} = 1 & \text{for } |x| \leqslant L_x/2 \text{ and } |y| \leqslant L_y/2, \\ = 0 & \text{elsewhere.} \end{cases} \tag{2.42}$$

Solving (2.27) for a large aperture ($L_x, L_y \gg 1$) we get the Fourier transform of the grading

$$f(l,m) = L_x L_y \frac{\sin(\pi L_x l)}{\pi L_x l}\cdot\frac{\sin(\pi L_y m)}{\pi L_y m} \tag{2.43}$$

and the field pattern follows from 2.28

$$F(l,m) = F_e(l,m) \, . \, \frac{\sin(\pi L_x l)}{\pi L_x l} \, . \, \frac{\sin(\pi L_y m)}{\pi L_y m}$$

$$\approx \frac{\sin(\pi L_x l)}{\pi L_x l} \, . \, \frac{\sin(\pi L_y m)}{\pi L_y m} \quad (\text{for } L_x, L_y \gg 1). \qquad (2.44)$$

The element pattern changes very little over the 'narrow' pattern of a large two-dimensional antenna and can therefore in most cases be treated as a constant. Solving 2.29 to 2.31 for the effective area we get

$$A(l,m) = \eta_R \lambda^2 L_x L_y \left\{ \frac{\sin(\pi L_x l)}{\pi L_x l} \, . \, \frac{\sin(\pi L_y m)}{\pi L_y m} \right\}^2. \qquad (2.45)$$

Note that the maximum effective area is equal to the geometrical area ($\lambda^2 L_x L_y$ m^2) for an ideal antenna with no internal losses, i.e. when the efficiency factor $\eta_R = 1$. The field pattern and the effective area of the uniformly graded rectangular aperture are shown in Figure 2.5.

2.4.3. *The circular ring and the uniformly graded circular aperture*

The Fourier transform is normally expressed in rectangular coordinates. As we might expect, similar transformations are possible when the coordinates are not rectangular. For circularly symmetrical systems the transforms are known as *Hankel transforms*. If the grading of an aperture is circularly symmetrical, then it is a function of the radius $\rho = (x^2+y^2)^{\frac{1}{2}}$, measured from the (x, y) origin. The field pattern will also be circularly symmetrical and we need know only the variation along one coordinate. For simplicity we make the calculations in the plane $l = 0$.

The two-dimensional Fourier transform is

$$f(l,m) = \iint_{-\infty}^{+\infty} g(x,y) \, . \, \exp\{j2\pi(xl+ym)\} \, dx \, dy. \qquad (2.27)$$

We put $l = 0$ and write $g(\rho)$ for the circularly symmetrical grading. Changing variables in the integral ($x = \rho \sin\phi, y = \rho\cos\phi$) (see Figure 2.6) we get

$$f(0,m) = \int_0^{2\pi} \int_0^{\infty} g(\rho) \, . \, \exp\{j2\pi\rho m \cos\phi\} \, \rho \, d\rho \, d\phi. \qquad (2.46)$$

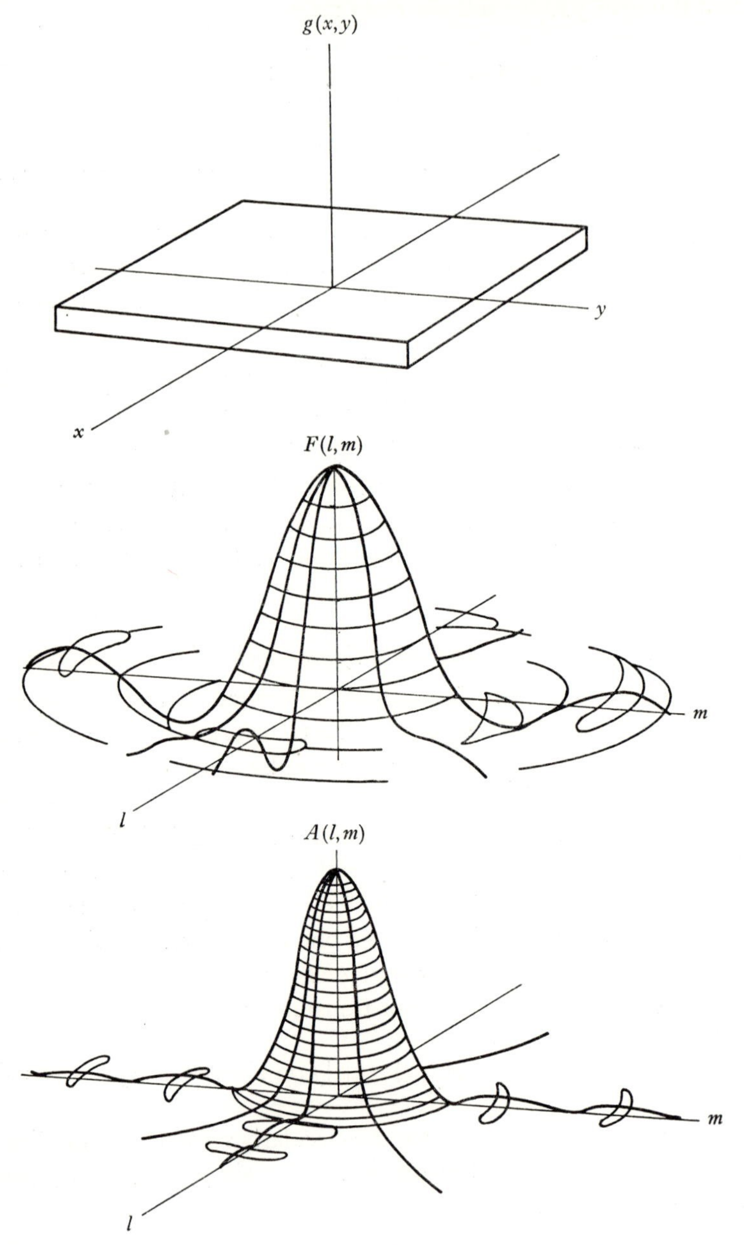

Figure 2.5. A uniformly graded rectangular aperture: the grading $g(x, y)$, the field pattern $F(l, m)$ and the effective area $A(l, m)$.

For *a thin ring of width dρ and radius* ρ_1 we have, with $g(\rho_1) = 1$,

$$f(0, m) = \rho_1 d\rho \int_0^{2\pi} \exp\{j2\pi\rho_1 m \cos\phi\}\, d\phi, \tag{2.47}$$

$$= 2\pi\rho_1 d\rho \,.\, J_0(2\pi\rho_1 m) \tag{2.48}$$

which expresses $f(0, m)$ in terms of a Bessel function of zero order.

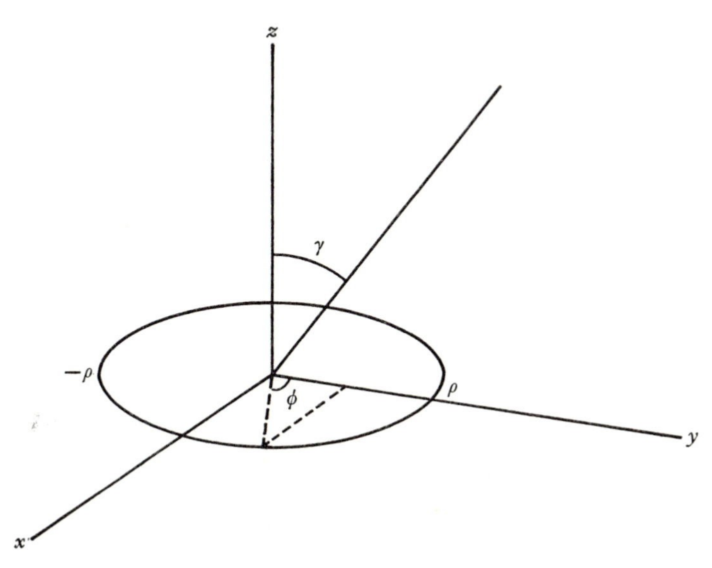

Figure 2.6. Coordinates used in calculating the field pattern of a ring antenna.

[Equation 2.48 is put in more familiar form if we replace m by $\sin\gamma$, where γ is the angle measured from the z-axis.] The extension of (2.48) to a circular aperture is straightforward. We have from (2.46)

$$f(0, m) = 2\int_0^{\infty} \rho g(\rho) \left[\int_0^{2\pi} \exp\{j2\pi\rho m \cos\phi\} d\phi\right] d\rho$$

$$= 2\pi \int_0^{\infty} \rho g(\rho)\, J_0(2\pi\rho m)\, d\rho, \tag{2.49}$$

where $f(0, m)$ is called the Hankel transform of $g(\rho)$.

The uniform illumination means that

$$g(\rho)\left\{\begin{array}{ll} = 1 & \text{for} \quad \rho \leqslant a, \\ = 0 & \text{for} \quad \rho > a, \end{array}\right\} \tag{2.50}$$

where a (wavelengths) is the half-diameter of the aperture.

The solution of (2.49) is

$$f(0, m) = 2\pi a^2 J_1(2\pi am)/(2\pi am). \tag{2.51}$$

Ignoring the slow variation of the element field pattern F_e and normalizing to unity in the maximum direction, we get the field pattern in the plane $l = 0$:

$$F(0, m) = F(\sin\gamma) = 2J_1(U)/U, \tag{2.52}$$

where $$U = 2\pi am = 2\pi a \sin\gamma$$

which has a similar form, with the Bessel function replacing a sine function, to the field pattern of a uniformly graded line antenna (2.37).

Equations 2.29 to 2.31 give

$$A(l, m) = A_{\max} \cdot |F(l, m)|^2 = \eta_R \pi a^2 \lambda^2 \{2J_1(U)/U\}^2. \tag{2.53}$$

The maximum effective area is equal to the geometrical area $\pi a^2 \lambda^2 (\mathrm{m}^2)$ when there are no internal losses, i.e. when $\eta_R = 1$. This was true also for the square aperture (2.45) and it is, in fact, a general property of uniformly graded large antennas. The relation between $A_{\max}$ and the shape of the grading will be discussed in more detail in Chapter 3, section 3.1.

2.4.4. *Antennas with non-uniform grading*

The current grading in some types of antennas is unavoidably non-uniform and in others, where it is easily adjusted, a non-uniform grading is often preferred, as this can reduce the level of unwanted responses (sidelobes) in the pattern. Any non-uniformity in the grading has the effect of reducing $A_{\max}$ while increasing the response in other directions so as to keep the all-sky integral of the effective area constant. These effects are shown qualitatively in Figure 2.7 for a simple line antenna and will be discussed further in Chapter 3.

The Fourier integral in (2.23) or (2.27) can become very complicated when the grading is non-uniform and may have to be evaluated numerically. However, for many common types of non-uniform gradings the integral can be solved directly, using some of the many remarkable theorems from Fourier analysis, particularly the addition, the convolution and the shift theorem (Appendix 2). Most readers will be familiar with these and we shall confine ourselves here to a brief description of their use in antenna calculations.

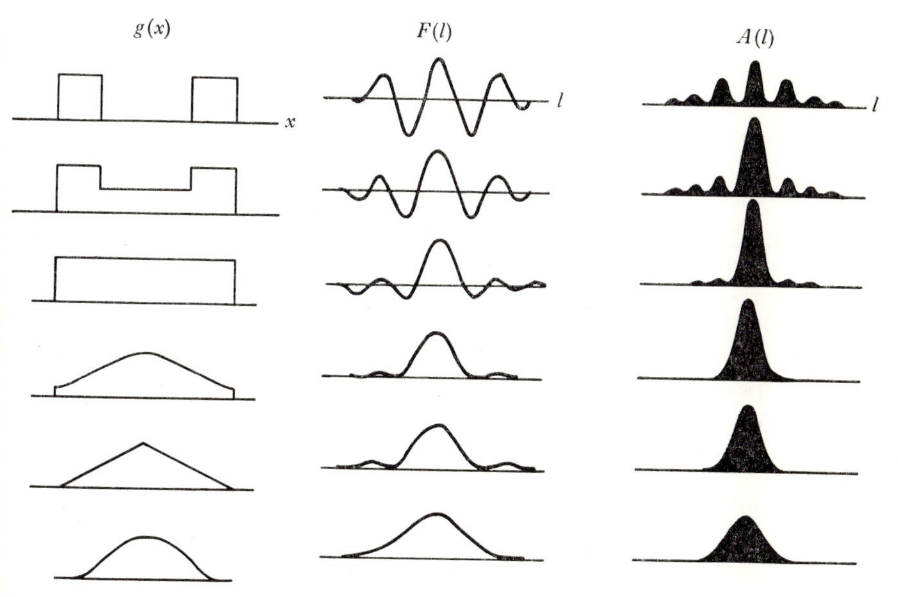

Figure 2.7. A selection of gradings $g(x)$ for a line antenna, the field patterns $F(l)$ and effective areas $A(l)$.

(*a*) *The addition theorem* tells us the almost self-evident fact that the Fourier transform of the sum of two gradings is the sum of the Fourier transforms of the gradings taken separately. We shall use this theorem also in the form that the Fourier transform of the difference of two gradings is the difference of the individual transforms.

(*b*) *The convolution theorem* is the most important of the theorems. It has been used in a limited form in antenna theory for a long time under the name of the 'array of arrays theorem'. Convolution is simple to visualize geometrically if we imagine the graph of one function to be slid over the graph of another (Figure 2.8). At any relative position x of the two, the integral of the product of the functions gives one point of a function called the convolution of the component functions. This is represented symbolically as

$$g_1(x) * g_2(x)$$

and is defined formally as

$$g_1(x) * g_2(x) \equiv \int_{-\infty}^{\infty} g_1(\sigma) g_2(x-\sigma)\, d\sigma. \tag{2.54}$$

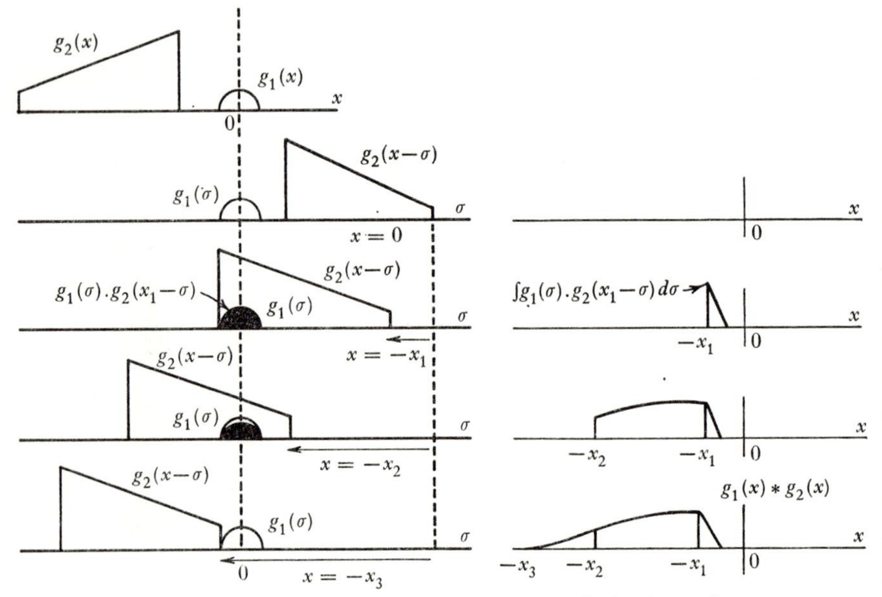

Figure 2.8. A geometrical representation of the convolution integral

$$g_1(x) * g_2(x) = \int_{-\infty}^{\infty} g_1(\sigma) g_2(x-\sigma) d\sigma.$$

At each separation x of the two functions the integral of the product of the functions gives one point on the convolution integral.

The convolution theorem states that

$$f_1(l) \cdot f_2(l) \mathscr{F} g_1(x) * g_2(x) \tag{2.55}$$

and

$$f_1(l) * f_2(l) \mathscr{F} g_1(x) \cdot g_2(x). \tag{2.56}$$

To apply this theorem to radio antennas we define $f_1(l)$ and $f_2(l)$ to be the Fourier transforms of the gradings $g_1(x)$ and $g_2(x)$. We can frequently express a complicated grading $g(x)$ as $g_1(x) . g_2(x)$ or as $g_1(x) * g_2(x)$, where $g_1(x)$ and $g_2(x)$ are relatively simple functions and (2.55) and (2.56) allow us to calculate the transform of $g(x)$. Figure 2.9 shows how a grading (*e*) which represents, for example, the one-dimensional grading of an array of separate paraboloids can be regarded as the product of an infinite array of spikes first multiplied by a finite uniform grading and then convolved with a

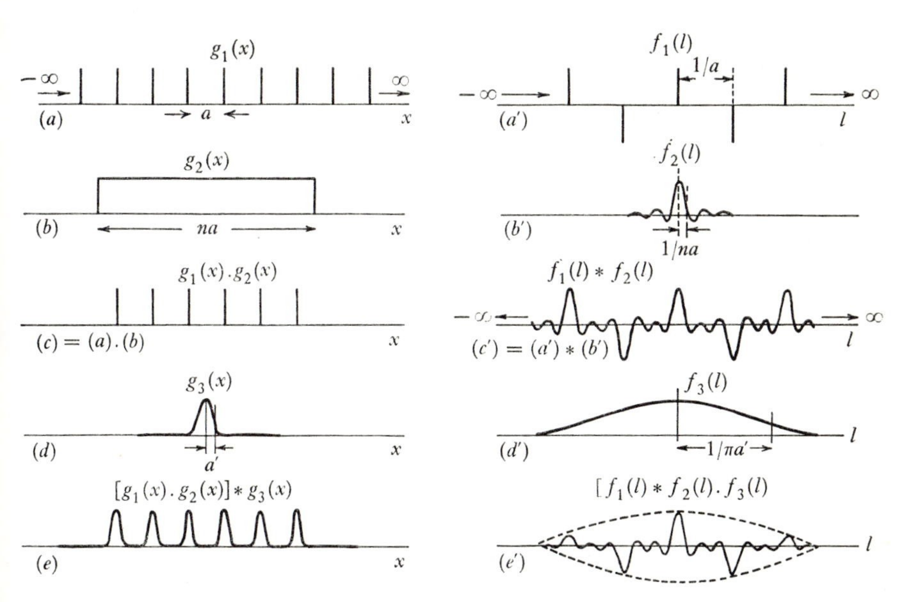

Figure 2.9. (*e*) represents the grading of a grating array of six paraboloids. (*a*), (*b*), (*c*), (*d*) show how this grading may be derived from the three gradings (*a*), (*b*) and (*d*), the transforms (*a*′), (*b*′) and (*d*′) of which are well known. The transform (*e*′) of (*e*) is then derived very simply from (*a*′), (*b*′) and (*d*′).

Gaussian function. The corresponding transforms are shown on the right-hand side of Figure 2.9. The calculation has been reduced to a series of very simple steps.

(*c*) *The shift theorem* can be applied to the electrical steering of an antenna. If we have the usual relation between $f(l)$ and $g(x)$

$$f(l) = \int_{-\infty}^{+\infty} g(x) . \exp\{j2\pi xl\} dx \qquad (2.23)$$

and introduce into $g(x)$ a linear phase shift of s radians per wavelength along the antenna so that $g(x)$ is changed into

$$g(x) \,.\, \exp\{j2\pi sx\},$$

then the transform has become

$$f_s(l) = \int_{-\infty}^{+\infty} g(x) \,.\, \exp\{j2\pi x(l+s)\}\, dx$$

$$\equiv f(l+s). \tag{2.57}$$

Equation (2.57), which is a statement of the shift theorem, means that a linear phase shift in $g(x)$ has introduced a shift in the directional cosine l—i.e. the antenna response has been steered to a new position.

CHAPTER 3

THE STEERABLE PARABOLIC REFLECTOR (PARABOLOID OF REVOLUTION)

The most commonly used of the radiotelescopes described in Chapter 1 is the steerable parabolic reflecting telescope, usually known as the parabolic dish antenna. It can be used singly as a simple antenna or as a unit of a large array type radiotelescope. In this chapter we shall be concerned only with the properties of the single parabolic dish.

The popularity of this type of antenna is the result of its electrical simplicity. Electromagnetic power received from a celestial source of radiation is brought to a focus at a 'point' after reflection from the paraboloidal surface and can be absorbed there by a simple horn or dipole 'feed' antenna. The reflecting surface is effective over a wide range of wavelengths, and observations may be made at many different wavelengths simultaneously. This is particularly valuable for radio spectroscopy.

The electrical simplicity of the parabolic reflector is matched, however, by its mechanical complexity. It must be possible to tilt and rotate it so that its axis can be pointed towards different parts of the sky and it must retain its shape in the process. It must also keep its shape under the influence of wind and temperature changes unless it is placed inside some transparent structure (radome) that shields it from such influences. The weight and cost of a rigid monolithic structure increases roughly as the cube of its linear dimensions, and large antennas of this type become very expensive.

The electrical performance of parabolic antennas cannot be calculated very accurately. Measurements are always needed to determine it precisely. Nevertheless, approximate calculations of performance can be made and, in fact, must be made for design purposes and in order to allow the measurements to be interpreted intelligently. Most of this chapter will be concerned with such calculations.

In section 2.3 we envisaged a situation in which all the currents on the antenna were transferred to an imaginary surface surrounding it. For simplicity, this surface was reduced to a plane in front of the antenna called the *aperture plane*. A further simplification was made in section 2.4.3 where the currents were assumed to be confined to an area of the aperture plane (*the aperture*) which corresponded to the projection of the reflecting surface. The field pattern of the antenna can be calculated without much difficulty if the phase, amplitude and polarization distributions of the currents over the transmitting aperture are known.

With a parabolic antenna we have to make some allowance for radiation from the feed antenna that does not pass through the aperture. This includes 'spillover' (feed radiation not intercepted by the reflector) and leakage through the reflector surface. A further difficulty is that usually we do not know with certainty the phase and amplitude (i.e. grading) of the current at all points in the aperture. So we usually make certain assumptions about the field pattern of the feed antenna and the effect of the reflecting surface on the radiation from the feed, and from these assumptions calculate approximately the current grading across the aperture. We must then make allowance for the disturbing effects of the feed and its supporting structure on the aperture grading and for the radiation from the feed which does not pass through the aperture.

In this chapter we shall discuss the field patterns of circular apertures with different gradings, the effects of imperfections in the reflecting surface on the antenna characteristics, and the characteristics of the feed antenna.

3.1. The current grading across the aperture

The field pattern of a uniformly graded circular aperture is (section 2.4.3)

$$F(0, m) = F(\sin\gamma) = 2J_1(U)/U, \qquad 2.52$$

where

$$U = 2\pi a \sin\gamma$$

and a is the half-diameter of the aperture in wavelengths. From this the effective area was found to be

$$A(l, m) \equiv A(\sin\gamma) = \eta_R \pi a^2 \lambda^2 [2J_1(U)/U]^2. \qquad (2.53)$$

With non-uniform grading more complicated integrals are involved, particularly when the grading is not of a circularly symmetrical form. A circularly symmetrical grading will, however, often be a good approximation to the real situation and we have

$$f(0, m) = 2\pi \int_0^\infty \rho g(\rho)\, J_0(2\pi\rho m)\, d\rho \qquad (2.49)$$

$f(0, m)$ is the Hankel transform of $g(\rho)$.

Several of these transforms have been tabulated. A very useful family of functions whose transforms can be derived easily from tables is

$$g(\rho) = K + [1 - (\rho/a)^2]^p \qquad (3.1)$$

for $\rho \leqslant a$ and $g(\rho) \equiv 0$ for $\rho > a$. This function can be fitted to most gradings encountered in practice by adjusting the two parameters K and p (Figure 3.1).

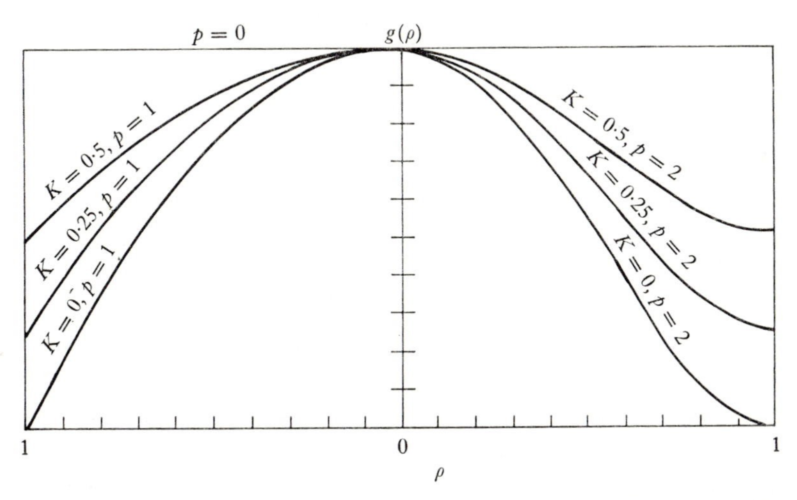

Figure 3.1. A family of useful gradings for a circular aperture. These are of the form

$$g(\rho) = K + [1 - (\rho/a)^2]^p.$$

The Hankel transforms of these can be derived easily from tabulated functions. Some of the beam characteristics corresponding to the various $g(\rho)$ plotted here are given in Table 3.1.

Antenna patterns are often characterized by a few numbers which describe their more important properties. The half-power *beam-width* is the angular distance between the directions at which the power pattern has fallen to 1/2—i.e. where the field pattern amplitude is down to $1/\sqrt{2}$. Other useful numbers are the angles at which the power pattern first falls to zero, the positions of its subsidiary maxima and minima, and its magnitude at the subsidiary maxima. The figures in Table 3.1 refer to an ideal aperture of diameter $D/\lambda(=2a)$ wavelengths and with gradings of the form given by (3.1). The more highly tapered gradings may be seen to produce greater beamwidths and smaller sidelobes than the more uniform gradings.

Table 3.1. *Characteristics of an ideal circular aperture of radius $a(=D/2\lambda)$ wavelengths which has a grading*

$$g(\rho) = K + [1-(\rho/a)^2]^p$$

p	K	Beamwidth between half-power points (radian)	Angular distance from maximum to first null (radian)	Amplitude first side-lobe (dD below main lobe)	$\frac{A_{max}}{\text{physical area}}$
0	0	$1.02\lambda/D$	$1.22\lambda/D$	17.6	1.00
1	0	$1.27\lambda/D$	$1.62\lambda/D$	24.7	0.75
2	0	$1.47\lambda/D$	$2.03\lambda/D$	30.7	0.55
1	0.25	$1.17\lambda/D$	$1.49\lambda/D$	23.7	0.87
2	0.25	$1.23\lambda/D$	$1.68\lambda/D$	32.3	0.81
1	0.5	$1.13\lambda/D$	$1.33\lambda/D$	22.0	0.92
2	0.5	$1.16\lambda/D$	$1.51\lambda/D$	26.5	0.88

The maximum effective area A_{max} depends upon the shape of the grading. It is equal to the physical aperture when the grading is uniform and there are no internal losses (2.53). A_{max} is always smaller than the physical aperture when the grading is non-uniform, as the last column of Table 3.1 suggests. An approximate formula for A_{max} for an ideal antenna without losses spillover, etc. can be derived from (2.17)

$$A_{max} = \lambda^2 \Big/ \int_{4\pi} P(l,m)\, d\Omega. \tag{2.17}$$

The main part of the reception pattern of any reasonably large paraboloid will be concentrated in a small region of the sky about the direction $l=0$, $m=0$. We can therefore write $d\Omega \simeq dl\,dm$ and replace the all-sky integral by an integral over l and m with the formal limits of $\pm\infty$. Further, the power pattern $P=|F|^2$ and, since the element pattern F_e will be nearly constant over this small region of sky, we have from (2.28)

$$F(l,m) \approx f(l,m)/|f(0,0)|. \tag{3.2}$$

The definition of a Fourier transform (2.27) shows directly that

$$f(0,0) = \iint g(x,y)\,dx\,dy. \tag{3.3}$$

Equation 2.17 can now be written

$$A_{\text{max}} = \frac{\lambda^2 \left| \iint g(x,y)\,dx\,dy \right|^2}{\iint |f(l,m)|^2\,dl\,dm}. \tag{3.4}$$

Parseval's (Rayleigh's) theorem in Fourier analysis states that

$$\iint |f(l,m)|^2\,dl\,dm = \iint |g(x,y)|^2\,dx\,dy$$

and we get

$$A_{\text{max}} = \frac{\lambda^2 \left| \iint g(x,y)\,dx\,dy \right|^2}{\iint |g(x,y)|^2\,dx\,dy}. \tag{3.5}$$

For a circularly symmetrical grading $g(\rho)$ expressed in polar coordinates ($a = D/2\lambda$) (3.5) takes the form

$$A_{\text{max}} = \frac{2\pi\lambda^2 \left| \int_0^a g(\rho)\,\rho\,d\rho \right|^2}{\int_0^a |g(\rho)|^2\,\rho\,d\rho} \tag{3.6}$$

which reduces to the physical area $\pi a^2 \lambda^2$ square metres for a uniform grading $g(\rho) = 1$ ($\rho \leqslant a$).

The physical meaning of these equations has been obscured by the mathematical argument. The distant field of the transmitting aperture in the axial direction is the sum of the effects of all the current elements. Hence the power flux in this direction, which is proportional to the square of the field, is proportional to the numerators of the expressions in (3.5) and (3.6). On the other hand, the total power flux through the aperture is the sum of the power flowing through all the elements of area of the aperture, and this is proportional to the denominator in the expressions (3.5) and (3.6). Hence (3.5) and (3.6) are proportional to the ratio of the power flux in the forward direction to the total transmitted power. This ratio must be proportional to the effective area in the forward direction, as can be seen from the reciprocity theorem and the definition of the effective area.

Internal losses in the antenna will, as usual, reduce the effective area as given by (3.4) and (3.5); the calculated A_{max} must be multiplied by the efficiency factor η_R. In a reflector type antenna there are three further factors that reduce the maximum effective area: spillover, transmission through the reflecting surface, and obstructions such as feed supports in front of the aperture. The first two may be looked upon simply as a loss of power in the forward direction when the antenna is transmitting. A certain fraction β of the total power transmitted from the feed is not reflected but passes outside the edge of the reflector or leaks through the surface. The maximum effective area as given by (3.4) and (3.5) must, therefore, be multiplied by $\eta_R(1-\beta)$ to take into account internal losses, spillover and transmission through the reflecting surface.

The effect of an aperture obstruction is twofold: it reduces the clear aperture, and it also causes a reduction in the total power transmitted from the aperture, since the part of the power from the feed that would otherwise pass through the obstructed area is scattered or absorbed. The reduction of A_{max} caused by a small obstruction in front of a uniformly graded aperture is, therefore, twice as large as might be expected from the simple reduction in the clear aperture. The effect is even more serious if (as with the feed assembly in a normal paraboloid, for example) the grading is non-uniform and the obstruction is located above the part of the aperture which carries the maximum current.

The obstruction can be thought of as producing a gap in the aperture. If this gap is large compared with the wavelength, we can neglect diffraction effects at the edges; the far field from the blocked aperture in the forward (maximum) direction will, then, be equal to the difference $(E-E')$ between the far field of the unobstructed aperture and the contribution to this far field that is due to the blocked-out portion. Power flux is proportional to the square of the field strength and we find that A_{max}, as calculated for the unobstructed aperture, must be multiplied by a factor of q in order to account for the effects of the obstruction where

$$q = (1 - E'/E)^2. \tag{3.7}$$

For a parabolic reflector with a circularly symmetrical grading $g(\rho)$ and diameter $2a$ wavelengths we have

$$E = \text{const.} \int_0^a g(\rho)\,\rho\,d\rho. \tag{3.8}$$

Let the obstructed region be a circular area $2a'$ wavelengths at the centre of the aperture. Then

$$E' = \text{const.} \int_0^{a'} g(\rho)\,\rho\,d\rho. \tag{3.9}$$

The fields are directly proportional to the aperture areas involved when the grading is uniform, i.e. when $g(\rho) = 1$. Equation (3.7) shows that A_{max} will be reduced to a quarter of its previous value if half of the aperture is obstructed. A small obstruction covering, say, 5 per cent of a uniformly graded aperture will reduce A_{max} by about 10 per cent.

The field pattern is given by the Fourier (or Hankel) transform of the blocked aperture grading. This is best calculated as the difference between the transforms of the unobstructed grading and of the blocked part (see the addition theorem, Appendix 2). Hence the blocking will add a wide low-amplitude negative component to the transform, causing the main field pattern maximum to be surrounded by a negative, i.e. phase reversed, wide and shallow, field sidelobe. The ratio of the solid angles occupied by the sidelobe and the main pattern is approximately equal to the ratio of the clear to the blocked part of the aperture. The power pattern shows the same sidelobe but, of course, positive and with an amplitude equal to the square of the field pattern amplitude, or approximately $E'^2/(E-E')^2$.

The supports of the antenna feed are, as a rule, more important in altering the illumination of the aperture by blocking or scattering than is the feed itself. The feed is usually supported by one, two, three, or four struts or legs. When one or two struts are used, then tensioning supports such as stay-wires may be used as well. The arrangements are shown in Figure 3.2. All the supports shown in this drawing will scatter some of the incoming radiation and block part of the aperture.

The arrangements (*a*) and (*d*) have quadrantal symmetry, an advantage when the antenna is used for polarization measurements. The single support 3.2*a*, guyed or unguyed, has some mechanical advantages, such as ease of access and of electrical connection to the feed. The disadvantage is that the support is in the region of strongest response for the feed. On the whole, it is preferable to have the feed situated closer to the mirror than the apex of its supports, so that there are no obstructions in its immediate vicinity.

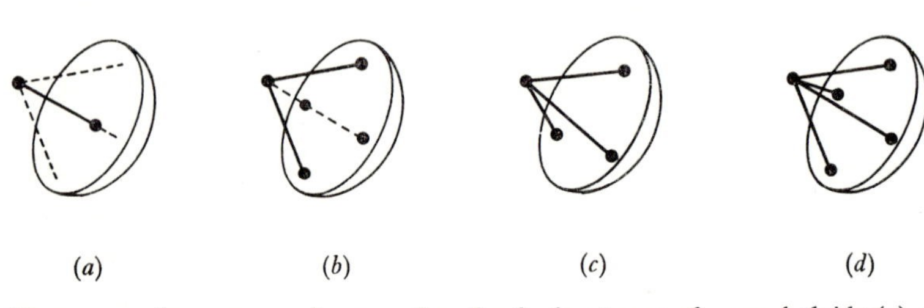

Figure 3.2. Some ways of supporting the feed antenna of a paraboloid: (*a*) a single central pole which may be guyed, (*b*) a pair of supporting legs in one plane with wire guys in the plane normal to this, (*c*) a tripod support, (*d*) a quadripod support.

It is obvious that the supporting legs should be as small in the cross-section as possible. No major improvement has been found in replacing the metal legs by non-metal ones. For the same mechanical strength the amount of scattering by steel and non-metallic legs appears to be comparable. Figure 3.3 shows the calculated and measured[(301)] disturbing effect of tripod legs on the aperture illumination.

3.2. The reflecting surface

The reflecting surface of the paraboloid must fulfil three conditions:

1. It should not depart significantly from the ideal shape in any conditions in which the telescope is used;

2. It should reflect practically all of the radio waves incident on it over the desired range of wavelengths and angles of polarization;

3. It should be as light as possible and offer the least wind resistance.

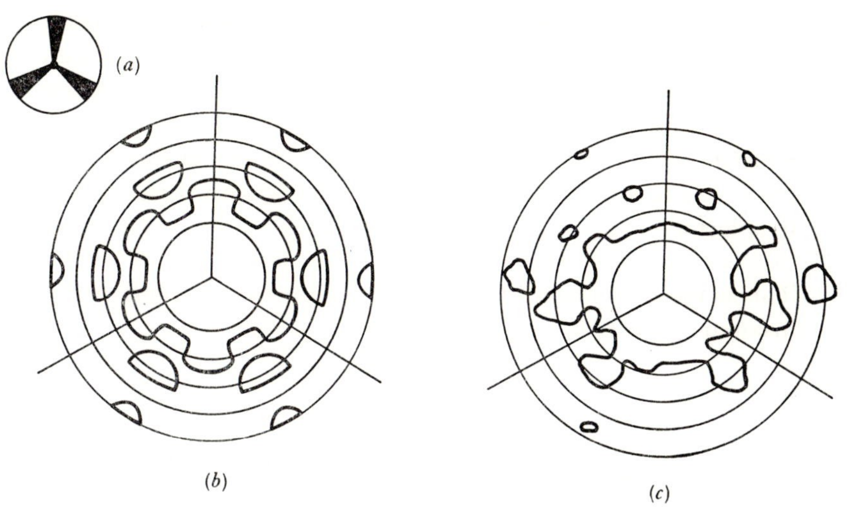

Figure 3.3. (*a*) A circular aperture with obscured sectors. (*b*) Some contours of the theoretical field pattern produced by the partly blocked aperture (*a*). 26 db contour. (*c*) Measured field contours of a paraboloidal antenna with the feed supported by tripod legs.(301) 30 db contour.

3.2.1. *Some effects of departure from ideal shape*

The departure of a reflecting surface from its ideal paraboloidal form depends, first, on errors in the construction and on flexure of the supporting structure. Secondly, the reflecting surface panels themselves can produce errors in phase, since they may depart from the ideal shape between the points at which they are attached to the structure. Frequently, for example, the reflecting surface is plane between the supports rather than curved.

The effect of phase errors in the aperture of a telescope has been considered by several writers.(302–5). Here we shall establish the magnitude of the effects only from elementary considerations. Let the surface of the reflector be indented at intervals not smaller than, say, a wavelength, so that the average departure from the true shape is $\pm\epsilon/\lambda$ wavelengths. These indentations will produce in the aperture (phase error doubled by reflection) an average phase error $\delta\phi = \pm 4\pi\epsilon/\lambda$ radians. We shall treat the antenna as if it were transmitting. The power projected from the feed onto the parabolic reflecting surface is the same, irrespective of whether the indentations are present or not. If there are no indentations, all reflections are additive in some particular direction, and at some distant point in this direction there is produced a field of intensity E (Figure 3.4*a*). For the indented antenna, half the aperture has an average phase error of $+\delta\phi$ and half of $-\delta\phi$ radians. If we combine these we find a distant field of $E\cos\delta\phi$. Hence, the radiated flux at this distant point will be lower than that from an antenna with perfect shape by a factor

$$\cos^2\delta\phi \approx 1-\delta\phi^2. \quad (3.10)$$

This is the ratio of the effective areas of the two antennas in the maximum direction.

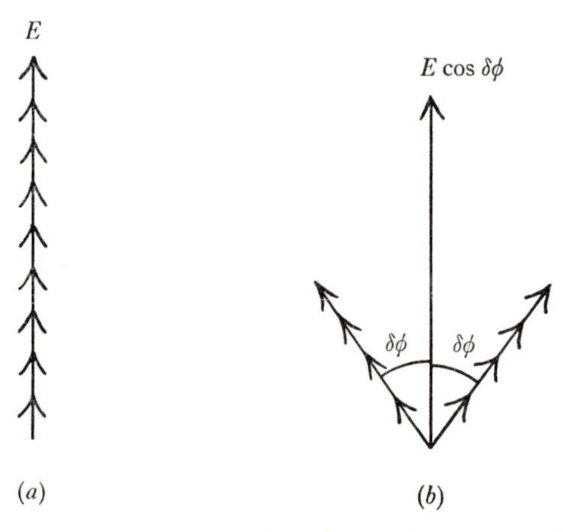

Figure 3.4. (*a*) Addition at a long distance of elements of the field from a uniformly phased aperture. (*b*) The resultant field if half of the elements in the aperture have their phase changed by $\delta\phi$ and half by $-\delta\phi$, i.e. the 'r.m.s.' error is $\delta\phi$ radians.

From such elementary reasoning we can deduce also the effect of the phase irregularities on the sidelobe response. The fields $E/2$ radiated from the two parts of the antennas, as well as having components $E/2 \,.\, \cos\delta\phi$, have components in quadrature phase, of magnitude $\pm E/2 \,.\, \sin\delta\phi$, or approximately $\pm E/2 \,.\, \delta\phi$ (Figure 3.4). These components cancel in the direction of maximum response but not necessarily in other directions. Imagine N such field components which are due to N phase errors, each of magnitude $E\delta\phi/N$ distributed over the whole aperture. If the regions have some regular arrangement, then it is possible that in *some* particular direction their effects will be additive and will produce a field

$NE\,\delta\phi/N$ or $E\delta\phi$. Hence the maximum possible sidelobe level produced by the phase errors will (in terms of power-flux) be $(E\,\delta\phi/E)^2$ or $\delta\phi^2$. Often, however, the errors are not regular but are randomly distributed. When this is so there is no direction where the fields from all the components are additive but, as is well known, the *power* contributions are additive. Hence, the power-flux in any direction produced by these random contributions is proportional to $N(E\,\delta\phi/N)^2$. Comparing this with the power-flux in the main lobe, which is proportional to $E^2(1-\delta\phi^2)$, we find that the sidelobe level is

$$\approx \delta\phi^2/N\,.\,(1-\delta\phi^2) \qquad (3.11)$$

compared with the main response. If $\delta\phi \ll 1$, this becomes

$$\approx \delta\phi^2/N. \qquad (3.12)$$

It should be noted that the amplitude of individual sidelobe responses depends on the total number of independent large-scale ($>\lambda$) irregularities in the antenna, whereas the loss of power in the main response (3.10) does not depend on this. Hence we must deduce that the total power radiated into the sidelobes is also independent (at least to the first order) of the number of the irregularities. It appears, therefore, that by increasing the number of independent irregularities we reduce the intensity of the sidelobes but increase their angular extent. When the size of the regions is reduced to a fraction of a wavelength, however, the error contributions produce only evanescent waves which have no effect at great distances. For this reason the irregularities involved in the use of a wire mesh-reflector surface can be disregarded in the range of wavelengths for which it is an efficient reflector.

A more careful study of the effects of phase irregularities in the aperture plane of a reflecting antenna requires a knowledge of the error distribution. In one such study [302] an assumption is made that errors are random in amplitude and distribution. A correlation distance C is defined as the distance at which, on average, the errors become essentially independent. At correlation distances, large compared with a wavelength, the theory gives results, both for the reduction of the main beam response and the sidelobe level, which are very close to those that we have derived from elementary reasoning. The effect on the main-beam response for different values of the correlation distance C is shown in Figure 3.5.

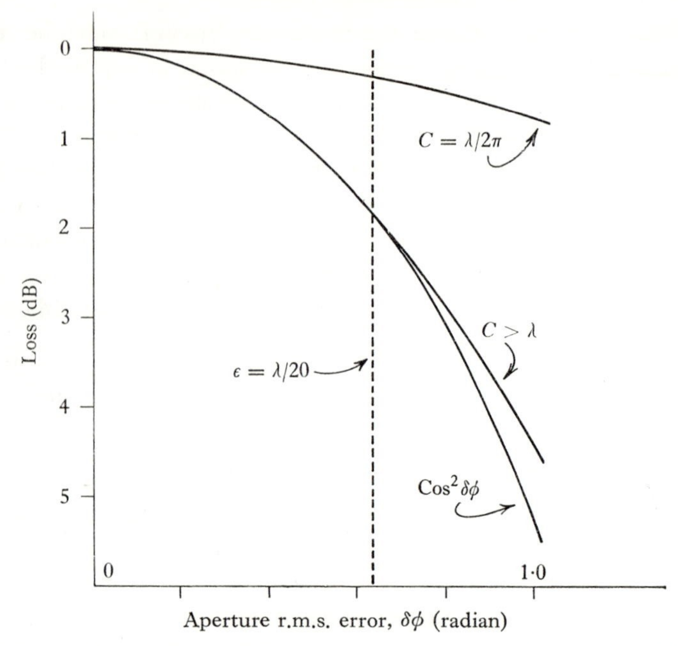

Figure 3.5. The loss in maximum effective aperture of a telescope caused by an r.m.s. phase error $\delta\phi$ across the aperture.

The elementary calculation given in the text is labelled $\cos^2\delta\phi$. The other curves were calculated by J. Ruze. For these C is the 'correlation distance' between irregularities.[302]

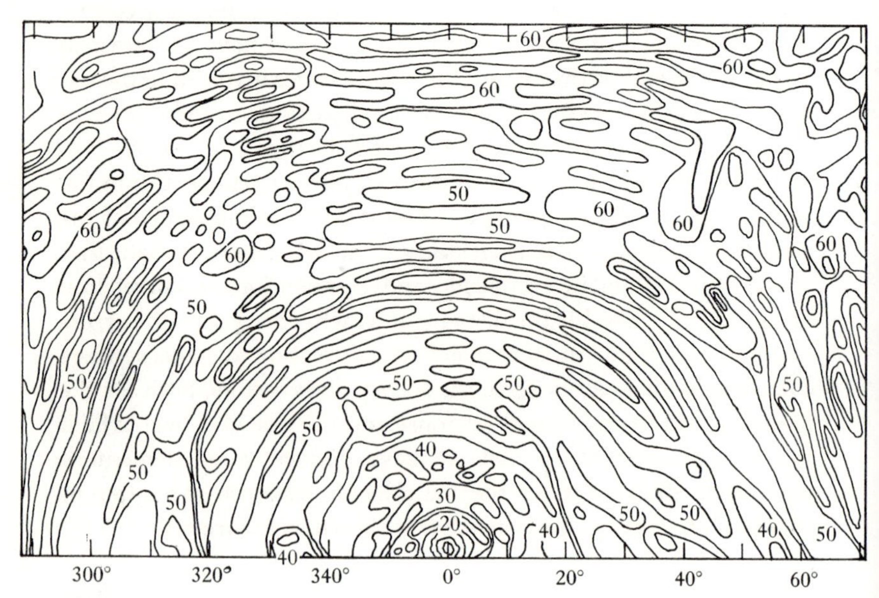

Figure 3.6. Sidelobe contours (in decibels below the maximum response) for a 25-metre diameter paraboloid operating at λ = 75 cm. (From C. A. Muller.)

Figure 3.6 shows the measured sidelobe level of a well-designed parabolic antenna.

3.2.2. *The effects of structural deflections on reflector performance*

The supporting structure of the reflecting surface of a steerable radiotelescope as well as the surface itself is subject to changing gravitational and wind forces and to changes in temperature.

For a family of scale models the gravitational forces increase roughly as the cube of the linear dimensions of the structure, i.e. as D^3, and the bending-moments as D^4. The wind forces, which are variable in magnitude as well as direction, are more difficult to specify, but for the purpose of crude estimations we can say that these, like the gravitational forces, are also proportional to D^3, i.e. the cross-sectional area D^2 multiplied by a wind pressure which we may take to increase in proportion to D, the height of the structure above ground. At high wind velocities the wind forces are usually much greater than the gravitational forces on the light structure of a radiotelescope and it may not be possible, therefore, to use the telescope in strong winds.

From elementary considerations it may be seen that the deflection in a rigid beam of fixed shape under the action of bending-moments proportional to D^4 will increase in proportion to D^2, i.e. as the square of its linear dimensions. Such deflections in the structure will produce deviations from the designed shape of the reflecting surface of the telescope; these deviations, moreover, will change in magnitude as the antenna is directed towards different parts of the sky. The deflections produced when one large telescope is shifted through 60° are shown in Figure 3.7.[306] When the allowable deterioration in performance of the reflecting surface has been specified, then (3.10) gives the tolerable r.m.s. phase error, $\delta\phi$; this corresponds to an r.m.s. surface irregularity ϵ (near the centre of the aperture), given by

$$\epsilon = \lambda_{min}\delta\phi/4\pi. \tag{3.13}$$

Frequently $\delta\phi$ is specified as being equal to 0·63 radians, making $\epsilon = \lambda_{min}/20$, although this involves a significant deterioration in the telescope performance (Figure 3·5). The minimum wavelength λ_{min} at which the telescope can be used, therefore, is given by 3.13. Taking $\delta\phi = 0{\cdot}63$, we have the minimum beamwidth of the antenna response given by

$$\text{Minimum beamwidth} \approx \lambda_{\min}/D = 20\epsilon/D. \tag{3.14}$$

We found earlier that the deflections caused by weight and wind pressure are roughly proportional to D^2. If we call these deflections ϵ then $\epsilon \propto D^2$ and

$$\text{Minimum beamwidth} \propto D. \tag{3.15}$$

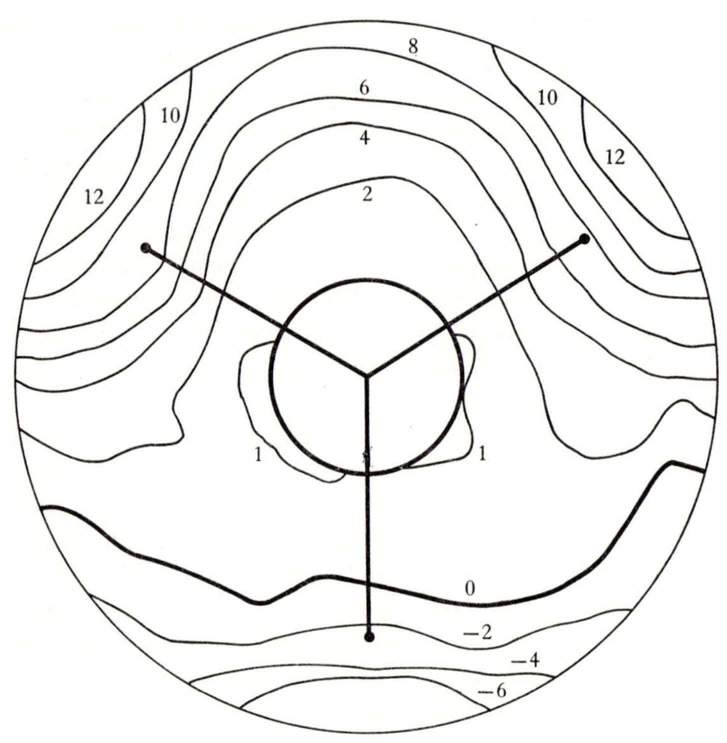

Figure 3.7. Structural deformations in millimetres of a 64-metre diameter paraboloid when it is tilted from the zenith to an angle 60° from the zenith.[306]

Equation 3.15 indicates that, when the structural deformations determine the surface errors of the reflector, the maximum resolving power is not improved by an increase in D, but is worsened. That this is so can be seen in Table 3.2, where the minimum beamwidth is seen to increase (but not linearly) with D.

The figures quoted in Table 3.2 refer to well designed and carefully made instruments in use at the time when this book was

Table 3.2. *Approximate performances of some telescopes*

Telescope	Diameter D[m]	ϵ r.m.s. [m]	$\lambda_{min} = 20\epsilon$	λ_{min}/D	Minimum beamwidth $1{\cdot}2\lambda_{min}/D$
I Optical (USA)	5	$0{\cdot}5\times10^{-7}$	$[1\times10^{-6}]$*	$0{\cdot}2\times10^{-6}$	0·05 sec. arc
II Radio (USA)	5	$0{\cdot}8\times10^{-4}$	$1{\cdot}6\times10^{-3}$	$3{\cdot}2\times10^{-4}$	1·3 min. arc
III Radio (USSR)	22	4×10^{-4}	8×10^{-3}	$3{\cdot}6\times10^{-4}$	1·5 min. arc
IV Radio (USA)	36	12×10^{-4}	24×10^{-3}	$6{\cdot}5\times10^{-4}$	2·8 min. arc
V Radio (Aust.)	64	30×10^{-4}	60×10^{-3}	$9{\cdot}5\times10^{-4}$	4 min. arc

* This optical telescope is used for wavelengths about 10 times smaller than λ_{min} as defined above. The resolving power of a large optical telescope is in practice limited by atmospheric scintillation to about 0·1″ or worse.

written. That the limit of accuracy has not been reached in the radiotelescopes is seen in the vastly superior accuracy of the optical telescope. If the accuracy of the radiotelescopes could be brought into conformity with optical telescopes (and $\epsilon \propto D^2$) then a 100 m diameter radiotelescope might be constructed for use at a wavelength of 1 mm and have a resolving power of about 1 sec of arc. It is unlikely that this will ever be accomplished.

A considerable improvement in the design of radiotelescopes is possible, however, and may be expected to follow the introduction of structures which deflect, but either do not distort the reflecting surface to the same extent or, alternatively, distort the reflecting surface into a new paraboloidal shape which requires only a change in the position of the feed antenna to restore it to full efficiency. When such improvements have been made, the problem of thermal expansion remains and, while radio antennas are made of materials with a high coefficient of expansion and are subject to considerable variations in ambient temperature, these differential expansions will limit the surface accuracy.

The buckling of a metallic structure or surface under the effect of thermal gradients is very well known. Its effect on the surface accuracy can be much larger than might appear at first sight, as the following elementary discussion will show. In Figure 3.8 we have a rod or plate AB clamped at the ends. A and B are separated by a distance d_0. As a result of a temperature difference ΔT° between the rest of the structure and the plate AB the plate increases in length to $d = d_0(1+\alpha\Delta T)$, where α is the linear coefficient of expansion of the material. The plate takes on the bowed shape (assumed to be a circular arc) which departs from the former position of the plate by a distance ϵ'.

From the geometry of the system we have

$$d = 2R\theta, \quad d_0 = 2R\sin\theta \quad \text{and} \quad \epsilon' = R(1-\cos\theta)$$

and we find that
$$\epsilon' \approx 0{\cdot}6 d_0(\alpha\,.\,\Delta T)^{\frac{1}{2}}. \tag{3.16}$$

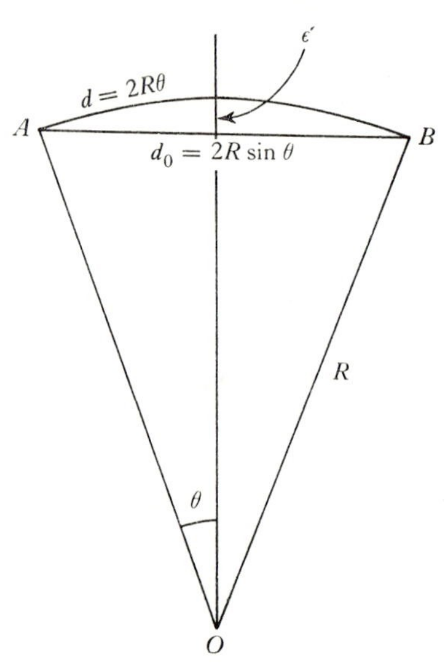

Figure 3.8. Bending of a flat plate AB when clamped at the ends and undergoing thermal expansion. R is the radius of curvature of the bend and ϵ' the deflection from the correct position.

If we make $d_0 = 1$ m, $\alpha = 10^{-5}$ per degree C, $\Delta T = 10\,^{\circ}$C, then $\epsilon' \approx 6$ mm, and if the surface takes up the corrugated form shown in Figure 3·8, the r.m.s. deviation $\epsilon \approx 4$ mm, which gives

$$\lambda_{\text{min}} \approx 8\text{ cm}.$$

This calculation shows why radiotelescopes can usually not be used at their shortest working wavelength when the structure is exposed to full sunlight.

3.2.3. *Reflectivity of the surface*

The reflecting surface of a radiotelescope may be continuous (metal sheet or sprayed metal) or it may be a discontinuous surface of spaced wires or rods. The discontinuous surface is generally cheaper and offers less resistance to wind than the continuous surface. Since the allowable gaps in the reflecting surface are proportional to the wavelength, the discontinuous structure has its greatest advantage at the longer wavelengths and the least at the shortest. It is never used at millimetre wavelengths, but is used invariably at metre wavelengths. The question that arises immediately with a continuous surface reflector is: How highly conducting must it be?

A simple approach to this question was demonstrated some years ago when it was shown that many problems concerning the transmission of plane waves may be treated as transmission line problems.[307] In this treatment the concept of 'intrinsic impedance of a medium' is used. This is analogous to the characteristic impedance of a transmission line. It represents the impedance presented to the wave; the magnitude of the impedance is referred to a square cross-section of the wavefront. For free space this is resistive and has a magnitude of 120π ($377\,\Omega$). It could be represented by a square non-conducting surface sprayed with conducting material to give $377\,\Omega$ between two opposite sides of the square (of any size). For a non-conducting medium other than a vacuum the value 120π is multiplied by $\sqrt{(\mu/K)}$, where μ and K are the relative permeability and relative permittivity of the medium. If plane waves are falling normally on a plane surface the problem may be represented as waves travelling along a transmission line meeting a discontinuity at which the characteristic impedance of the line suddenly changes. At the discontinuity, part of the wave will be reflected and part will continue along the second section of transmission line. If the second medium (or second section of transmission line) is a shunting conductor, then the continuing wave is rapidly damped out. With copper the attenuation is $130\nu^{\frac{1}{2}}$ decibels per metre and at a frequency of 10^8 Hz the wave will be attenuated by 13 decibels in one-thousandth of a centimetre.

Our main interest here, however, is in the fraction of the energy that is reflected at the discontinuity, not in how quickly the fraction that is transmitted is damped out. We need note only that this distance is extremely small and for any continuous metallic surface, even a thin layer of sprayed metal, the transmitted wave can usually be neglected.

For a conducting medium the intrinsic impedance is given by

$$Z_0 = \left[\frac{j\omega\mu}{\sigma + j\omega K}\right]^{\frac{1}{2}}, \tag{3.17}$$

where σ is the conductivity of the medium. For a metal this becomes

$$Z_0 = (\omega\mu/2\sigma)^{\frac{1}{2}}.(1+j). \tag{3.18}$$

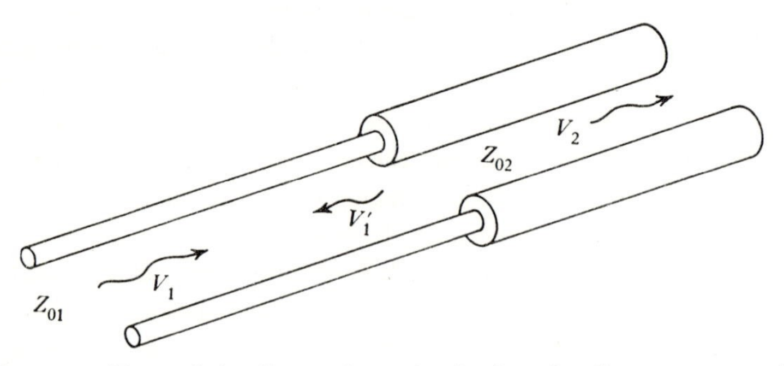

Figure 3.9. Transmission-line analogy of reflection of radio waves at an air-metal surface. V_1 is the voltage of the incident wave; V_2 refers to the transmitted wave, and V_1' to the reflected wave.

For copper $Z_0 \approx 2{\cdot}6\,.\,10^{-7}\nu^{\frac{1}{2}}(1+j)$ and $|Z_0| \approx 3{\cdot}7 \times 10^{-7}\nu^{\frac{1}{2}}$. If

$$\nu = 10^8\,\text{Hz}, \quad |Z_0| \approx 3{\cdot}7 \times 10^{-3}\,\text{ohm}.$$

Since the resistivity of the least conductive metals (bismuth, cast iron, resistive alloys) is not more than 100 times that of copper, the intrinsic impedance of all metals is less than on-tenth of an ohm at $\nu = 10^8$ Hz, or less than one ohm at $\nu = 10^{10}$ Hz.

In Figure 3·9 Z_{01} is the intrinsic impedance of free space Z_{02} refers to the reflecting medium.

From the elementary theory of propagation along transmission lines we know that the amplitudes of the incident, reflected and transmitted (or absorbed) waves (V_1, V_1', V_2) are given by

$$V_1/(Z_{02}+Z_{01}) = V_1'/(Z_{02}-Z_{01}) = V_2/2Z_{02}. \tag{3.19}$$

If we have a wave arriving from space normally to a metallic surface we have

$$Z_{01} = 120\pi \quad \text{and} \quad Z_{02} = R_{02}(1+j) = 2{\cdot}6 \, . \, 10^{-3}(1+j). \quad (3{\cdot}20)$$

For copper, at $\nu = 10^8$ Hz

$$\frac{|V_2|}{|V_1|} \approx 2 \times 10^{-5}, \quad \text{i.e. 94 db attenuation.}$$

For zinc at $\nu = 10^{11}$ Hz, the attenuation is 38 db. Continuous metal reflectors are highly efficient over the whole range of frequencies used in radio astronomy. The amount of energy absorbed or transmitted is negligible; hence, the amount of radio emission radiated thermally into the feed antenna by the metallic surface is also negligible.

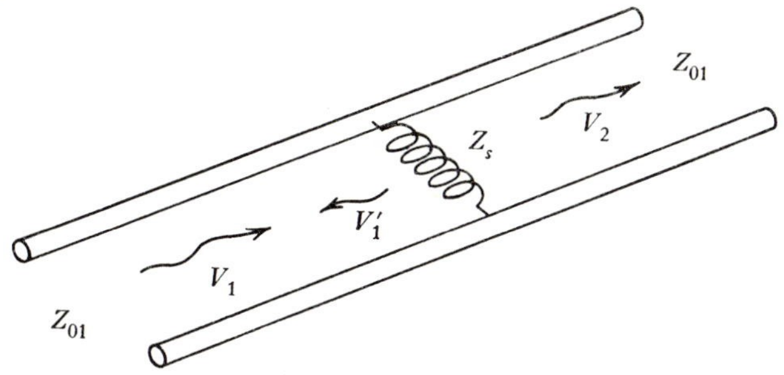

Figure 3.10. Transmission-line analogy of reflection at a wire grid of impedance Z_s.

3.2.4. *Mesh or grating reflectors*

We use a similar elementary treatment for a mesh reflector. A reflecting mesh or grating corresponds to a lumped impedance placed across the line at some point, as shown in Figure 3.10. We can use the relations (3.19) except that now

$$Z_{02} = Z_{01} \, . \, Z_s/(Z_{01}+Z_s). \quad (3{\cdot}21)$$

The amplitude of the transmitted wave compared with that of the incident wave is given by

$$V_2/V_1 = 2Z_{02}/(Z_{02}+Z_{01}) = 1/(1+Z_{01}/2Z_s). \quad (3{\cdot}22)$$

If we calculate the reactive impedance of a wire screen by the simple expression(307)

$$X_s = \frac{377 d \log_e [d/2\pi a]}{\lambda}. \quad (3{\cdot}23)$$

where $\qquad a =$ radius of wire

and $\qquad d =$ spacing between wire centres

we can get an approximate value for the attenuation of a signal through the system.

The surface of a parabolic dish is not normal to the wave front except in the centre, and the ratio Z_{01}/Z_s in (3.22) must be divided by the cosine of the angle between the incident wave front and the reflecting screen.

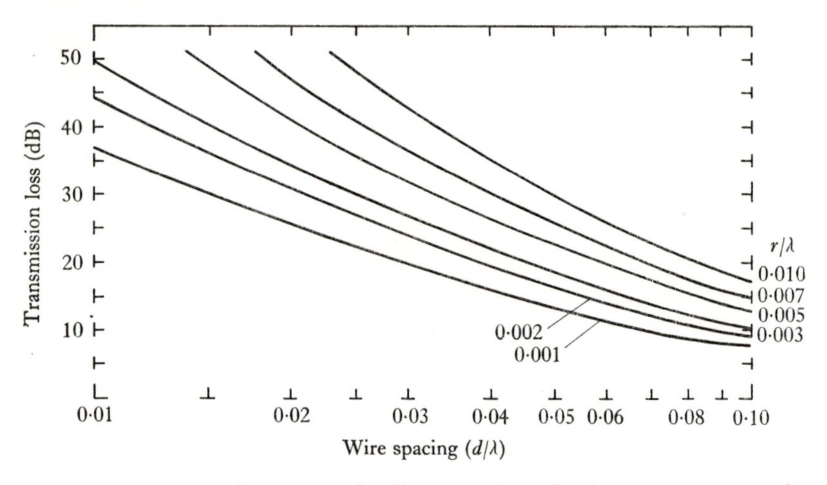

Figure 3.11. Transmission loss of radio waves through wire screens composed of wires of various radii (r/λ) as a function of the wire spacing (d/λ).

The resistive part of Z_s is usually negligible compared with the reactive part, X_s. For example, if the spacing between copper wires of radius 1 mm is 1 cm and the wavelength used is 30 cm, then $X_s \approx 6\,\Omega$. For the same screen R_s is approximately $0{\cdot}013\,\Omega$. The resistivity of the material of the wires could be increased by a factor of 1000 without affecting the efficiency of the reflecting screen—e.g. highly resistive stainless steel wire could be used.

With this value for X_s we have

$$V_2/V_1 \approx j/30 \quad \text{or an attenuation of about 30 db.}$$

This may be compared with the value given by a more exact treatment shown in Figure 3.11.[308]

This wire screen has high efficiency as a reflector, since little energy escapes through the screen. In some circumstances, however, an attenuation of 30 db might be inadequate to reduce sufficiently the interference from strong sources of radio emission behind the reflecting screen. Nevertheless, it must be remembered that the wanted signal has its effect increased by a factor equal to the gain of the antenna. This factor, which may be as high as 40 or 50 db, when added to the 30 db of attenuation through the screen, gives a discrimination against signals through the screen of 70 or 80 db.

We have confined the discussion to wire antennas in which the wires are arranged parallel to the direction of polarization of the

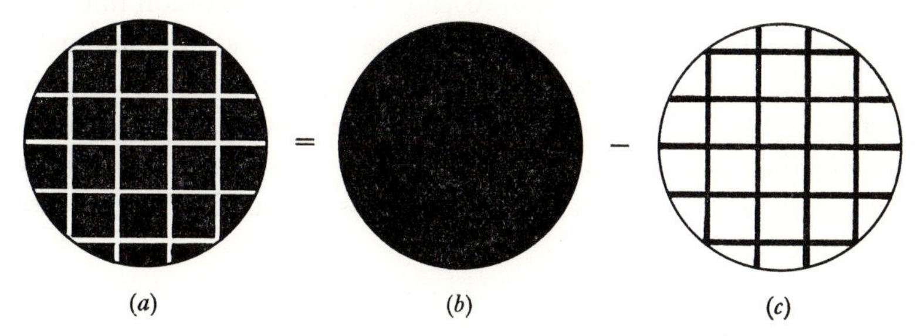

Figure 3.12. The current grading (*a*) in an aperture composed of separated panels is equivalent to the uniform grading (*b*) minus the grading (*c*).

waves. A reflector of this type will reflect only the component of the incoming radiation which is polarized in the direction of the wires. For a parabolic dish a wire mesh is suitable, as it has mechanical advantages and also reflects waves of all angles of polarization, so that polarization measurements may be made. It may be treated as two perpendicular sets of parallel wire screens.

Frequently the wire mesh is arranged in separate panels. The effect of such a division of the surface is negligible at wavelengths small compared with the linear dimensions of a panel. At long wavelengths, however, the presence of the gaps may reduce to some extent the efficiency of the reflecting surface. If we imagine the gaps in the surface to be wide enough to interrupt the current flow, as in Figure 3.12*a*, then by use of Babinet's principle we can regard the surface as equivalent to a continuous current sheet (Figure 3.12*b*) on which is superimposed a grid wherein the negative current distribution follows the dividing lines between the panels of the reflector (Figure 3.12*c*).[309] We can then say that, when the wavelength becomes great enough for the large grid of Figure 3.12*c* to become a reasonably efficient reflector, then the reflector (Figure 3.12*a*) becomes inefficient.

In practice the panels can be designed to have high capacitance between the edges of adjacent panels, and the effect of the gaps on the current distribution on the reflecting surface is not great at even the longest wavelengths at which the antenna is used.

3.3. The feed antenna

Energy is conveyed from the radio transmitter to the feed point, or from the feed point to a radio receiver, by a transmission line; the losses in this line may affect the performance of the system seriously. In receiving, the deterioration may be even more serious than in transmitting, because then we are concerned not only with attenuation of the signal but also with the *adding of noise* to the signal by the resistive components that produce this attenuation. We discuss this in Chapter 8.

3.3.1. *Prime-focus feed*

The feed antenna is required to illuminate the reflector in such a way as to produce a suitable distribution of phase and amplitude across the aperture. There is no 'ideal' distribution, but a distribution uniform in phase and amplitude is often aimed at, since this gives the greatest value of A_{max} for the antenna. Unwanted responses (sidelobes) for this distribution are not negligible but are small enough to cause little trouble in most observations.

Another requirement for the feed antenna is that it should not illuminate anything outside the angle subtended by the parabolic reflector—in other words, it should not produce any spillover. In receiving, this spillover not only wastes energy but causes radiation to be picked up from the relatively hot ($\sim$ 300 °K) ground or intense radio sources in the sky or sources of radio interference on the earth. To make matters worse, this unwanted pickup varies as the antenna is directed to different places in the sky. We shall return to this later.

The required polar diagram (power pattern) of the feed antenna depends, of course, on the focal ratio F/D (where F is the focal length and D the diameter of the aperture), and in existing paraboloids this may have any value between about 0·25 (with the focus in the aperture plane) up to about 0·6. Two factors which must be considered when selecting a focal ratio are the ease with which the reflector can be illuminated satisfactorily and the reduction of spillover. A focal-plane reflector ($F/D = 0{\cdot}25$) can be designed to have very small spillover, but it has an aperture illumination which is far from uniform. A long-focus reflector usually has these merits and faults reversed.

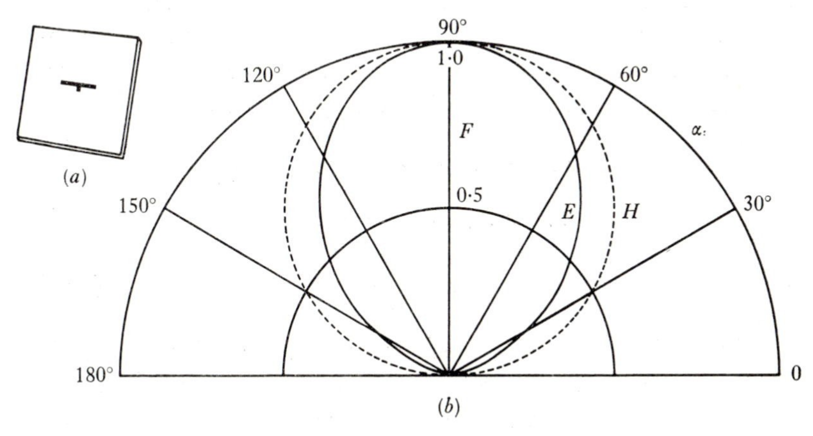

Figure 3.13. Field patterns in the E and H planes of a dipole placed parallel and close ($\sim \lambda/10$) to a reflecting plane. The angles marked represent α for the E curve and β for the H curve, when the dipole is parallel to the x axis. Note that $\gamma = 0°$ when $\alpha = \beta = 90°$.

Figure 3.13*a* shows the simplest type of feed antenna, a dipole placed a short distance in front of a large reflecting plate. Figure 3.13*b* shows the field pattern in the plane of the dipole (the 'E' plane) in a plane normal to the axis of the dipole (the 'H' plane).

A reflector which subtends an angle of ± 90 at the feed would intercept all the energy radiated from the latter if the polar diagrams were exactly as shown in Figure 3.13*b*. There are two factors, however, which prevent this. The first is that the field intensity at $\gamma = 90°$ is zero (except in the E plane) only when an *infinite* plane reflector is placed behind the dipole; the second is that the field at $\gamma = 90°$ falls to zero only at very long distances and not at the relatively short distance between feed and dish. With very large focal-plane parabolic reflectors hundreds of wavelengths in diameter fed by a dipole backed by a large plane reflector at the focus the situation approximates to that shown in Figure 3.13*b* and there is very little spillover of radiation.

With such a feed the illumination of the dish is far from ideal. It falls off rapidly with γ, not only because of the shape of the field pattern of the feed, but also because the distance r between the feed and the parabolic reflector increases with γ. This is shown in Figure 3.14, where it will be seen that the area δA of the aperture illuminated by a small solid angle $\delta\Omega$ is proportional to r^2.

The power flux per unit area of the aperture, therefore, is proportional to $(1/r)^2$ and the field intensity across the aperture shown in Figure 3.14 must be proportional to $1/r$. Now r can be found in terms of γ.

$$r+t = \text{constant} \quad \text{and} \quad t = r\cos\gamma,$$

$$\therefore \quad r(1+\cos\gamma) = \text{constant}$$

or

$$r \propto \frac{1}{1+\cos\gamma}. \tag{3.24}$$

The field pattern of a feed antenna such as that shown in Figure 3.13 must therefore be multiplied by a factor $(1+\cos\gamma)/2$ (normalized with respect to the centre) before it can represent the distribution over the aperture of the antenna. In Figure 3.15 this correction factor is applied to a $\lambda/2$ dipole with a plane reflector. The relation between the angle γ_{max}, which the edge of the reflector subtends at the feed, and the corresponding value of F/D (Figure 3.14) can easily be shown to be

$$F/D = (1+\cos\gamma_{max})/4\sin\gamma_{max}. \tag{3.25}$$

Figure 3.15 shows F/D ratios for $\gamma_{max} = 30°, 60°, 90°$.

It is obvious that with such a feed it is impossible both to avoid spillover and to illuminate the aperture evenly. At

$$F/D = 0.25 \quad (\gamma_{max} = 90°, \text{ a focal-plane aperture})$$

there is little spillover of energy, but the uneven illumination reduces the effective area of the aperture well below the value that it would have if evenly illuminated. We dealt with this in section 3.1.

In order to obtain both minimum spillover and even illumination

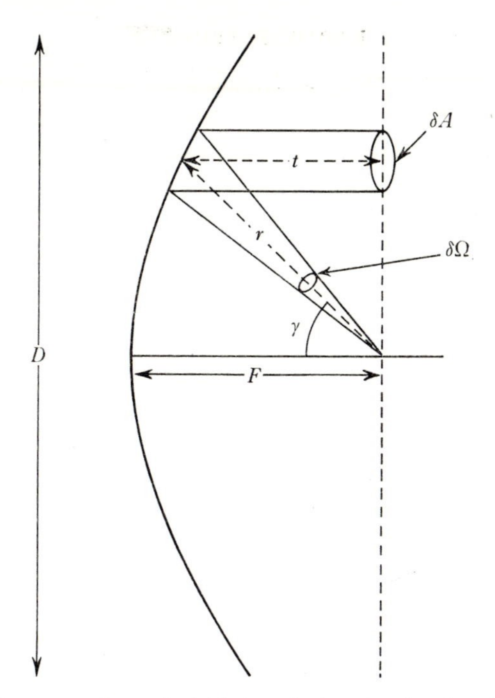

Figure 3.14. Geometry of a paraboloid of revolution showing the relation between a solid angle $\delta\Omega$ at the focus and the corresponding area δA of the aperture.

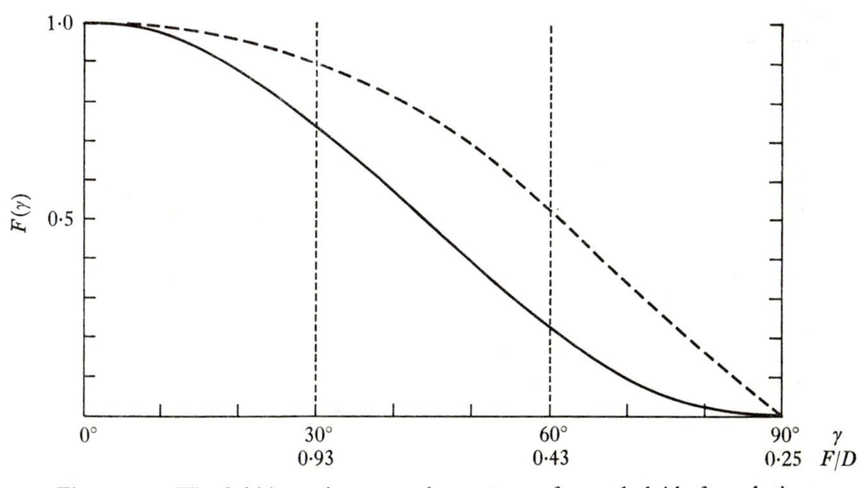

Figure 3.15. The field intensity across the aperture of a paraboloid of revolution when fed by a half-wave dipole backed by a plane reflector.

Two sections are shown, both through the axis of the paraboloid. One is parallel and the other perpendicular to the dipole. A maximum angle $\gamma = 90°$ corresponds to a focal plane feed, i.e. $F/D = 0{\cdot}25$. Maximum angles $\gamma = 60°$ and $\gamma = 30°$ correspond to $F/D = 0{\cdot}43$ and $F/D = 0{\cdot}93$ respectively. —, E plane; ----, H plane.

the polar diagram of the feed should be similar in all axial planes. The great difference between the patterns in the E and H planes for the simple dipole antenna with reflector, shown in Figures 3.13 and 3.15, is unsatisfactory. The difference can be reduced, for example, by the use of a horn or by an array of dipoles.

Figure 3.16*a* shows the field pattern of a feed composed of a dipole-pair and a plane-reflector.[310] Figure 3.16*b* shows that of a circular horn feed.[311] Each feed was designed to have similar characteristics in the E and H plane and to have minimum spillover, both of which are vitally important in measurements of polarization.

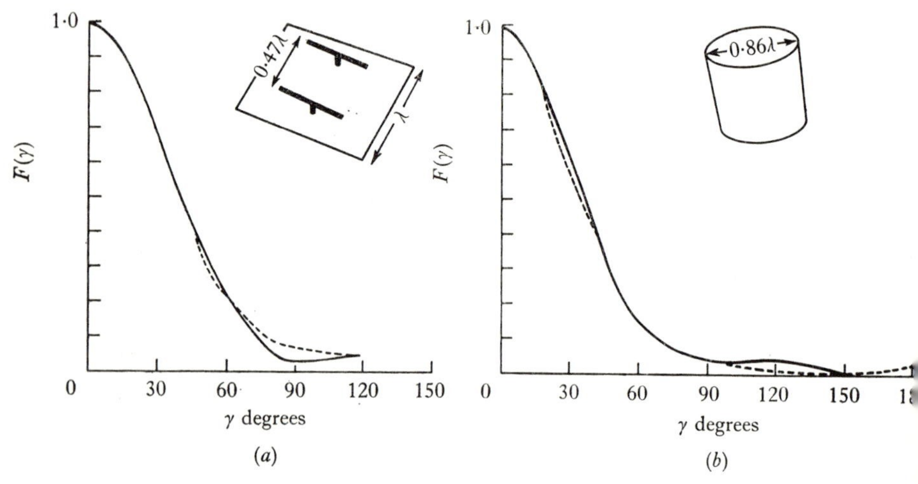

Figure 3.16. Two feeds designed to have similar field patterns $F(\gamma)$ in E (—) and H (----) planes. (*a*) A two-dipole array (E. Harting). (*b*) A circular horn (R. Wielebinski).

The feed antennas which we have described are of a very simple type. They produce a grading $g(\rho)$ across the aperture that is far from uniform or has a great amount of spillover; consequently the effective area of the aperture is markedly less than the physical area.

This can be viewed in another way. If we consider the arrival of plane waves at the aperture, then these waves, after reflection at the parabolic surface, converge at the focal plane and form the well-known diffraction pattern of Airy. The simple horn or dipole accepts only the energy in the central disk of this pattern; the rest is wasted. This is equivalent to the statement that the feed does not produce a uniform illumination or grading across the aperture. For uniform grading we must have a feed that collects all the energy from a plane wave that falls on the aperture, i.e. the feed must cover a large part (theoretically, all) of the Airy diffraction pattern and combine all the received energy in the correct phase to be fed to a receiver.

Suppose that we have a parabolic reflector of relatively long focal length (say $F/D > 0{\cdot}8$) and large diameter ($D \gg \lambda$), then the calculation of the current grading $g(\rho)$ at the focus required to produce a uniform grading $[g'(\rho') = 1]$ across the aperture of the antenna can be reduced to a problem of producing a field pattern of the feed given by

$$F_F(\sin\gamma) = \begin{cases} 1 & \gamma \leqslant \gamma_{\max} \\ 0 & \gamma > \gamma_{\max} \end{cases}.$$

Analogously to the inverse problem solved earlier (2.51 and 2.52) we have the solution (ignoring the variation of the element field pattern in this simple approximation)

$$g(\rho) \propto 2\pi \int_0^{\sin\gamma_{\max}} J_0(2\pi\rho\sin\gamma)\sin\gamma\,.\,d(\sin\gamma)$$

$$= 2\pi\sin^2\gamma_{\max} J_1(U')/U', \quad (3{\cdot}26)$$

where $U' = 2\pi\rho\sin\gamma_{\max}$. Antenna gradings are usually presented normalized to unity at the position of maximum current. Hence,

$$g(\rho) = 2J_1(U')/U'.$$

If we choose $\sin\gamma_{\max} = 0{\cdot}5$ (i.e. $F/D = 0{\cdot}93$), then $U' = \pi\rho$. The first minimum of the grading occurs at $\rho = 3{\cdot}85/\pi = 1{\cdot}23$ wavelengths (see Figure 3.17). This feed, of course, should extend to infinity, but we can in practice make a truncated feed which will approximate the first few oscillations of the function $2J_1(U')/U'$. Figure 3.18 shows the field pattern of a feed which consists of four current rings of relative amplitudes $1{\cdot}0$, $-0{\cdot}14$, $+0{\cdot}06$, $-0{\cdot}014$. The scale is for $\sin\gamma_{\max} = 0{\cdot}866$, i.e. $F/D = 0{\cdot}43$. With such a short focal length the optimum field pattern is not rectangular but has to include the distance factor $2/(1+\cos\gamma)$ discussed earlier. The optimum shape is

$$F_F(\sin\gamma) = \begin{cases} \dfrac{2}{1+\cos\gamma} & (\gamma \leqslant \gamma_{\max}), \\ 0 & (\gamma > \gamma_{\max}) \end{cases} \quad (3{\cdot}27)$$

as shown by the dotted line in Figure 3.18. The required function $g(\rho)$ differs somewhat from $2J_1(U')/U'$. In the truncated model of Figure 3.18 we can get a closer approximation to the required curve by decreasing slightly the amplitude of the current in the first ring with respect to the currents in the other rings.

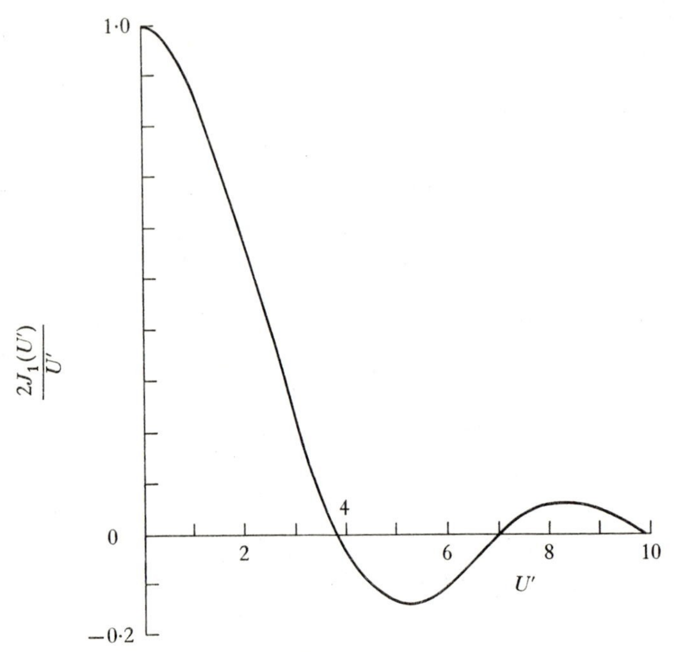

Figure 3.17. Normalized Fourier (Hankel) transform of a function that is uniform within a circular boundary and zero outside it. $J_1(U')$ is a first-order Bessel function.

Many more current rings would be required to produce a close approximation to the curve of (3.27) and the design problem for small values of F/D has not yet been solved. With large values of F/D this current grading can be produced by a secondary mirror.

3.3.2. *Multiple feeds and image formation*

In Chapter 2 we showed that a linear phase shift in the grading across an aperture is equivalent (apart from foreshortening the aperture) to turning the aperture through an angle. With a long focus ($F/D > 1$) parabolic reflector like that used in optical astronomy an image is formed in the focal plane of the region of the sky which lies within a small angular distance from the axis of the paraboloid. In reverse, this indicates that rays starting from a point in the focal plane near the geometrical focus will produce a grading which has a linear phase shift across the aperture. For regions of the sky further from the telescope axis a clear image is not formed and the image is said to suffer from 'coma' distortion. In optics this is

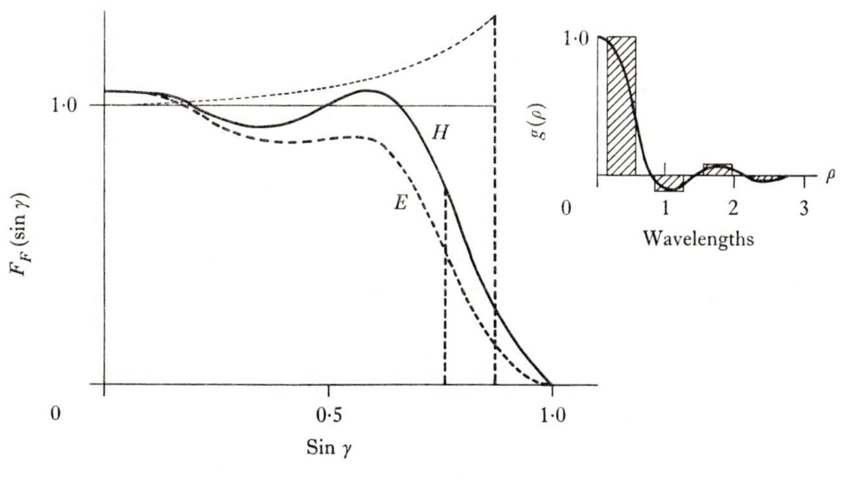

Figure 3.18. (*a*) A feed antenna composed of four rings of radiating elements with a grading that is a rough approximation to $(2J(U')/U')$ (smooth curve). (*b*) The field patterns $F_F(\sin\gamma)$ of the feed in the E and H planes.

The upper dotted curve is the ideal feed pattern (for uniform aperture illumination) for a paraboloid for which $F/D = 0{\cdot}43$ when the distance from the focus to different parts of the reflector is taken into account.

particularly troublesome with telescopes in which the ratio $F/D < 1$. We should expect, therefore, that coma, which indicates a non-linear phase shift in $g(\rho)$, would limit the image-forming capacity of radiotelescopes which usually have $F/D < 1$.

We shall now calculate the coma distortion in a simplified way in which we make rough approximations without justifying them. In Figure 3.19 the ray paths are drawn for a feed B' that is displaced a distance $BB' = \epsilon$ from the focus of the paraboloid. The path length from the focus B to the aperture plane OO' is equal to $2F$. The ray path from B' to the plane OO' differs from this by approximately

$\epsilon \sin i$. Then

$$\epsilon \sin i = \epsilon \,.\, EB/CB = \epsilon \,.\, \frac{4F\,.\,H}{4F^2+H^2}$$

$$\approx \delta\gamma\left(H - \frac{H^3}{4F^2} + \ldots\right), \qquad (3.28)$$

where $\delta\gamma = \epsilon/F$.

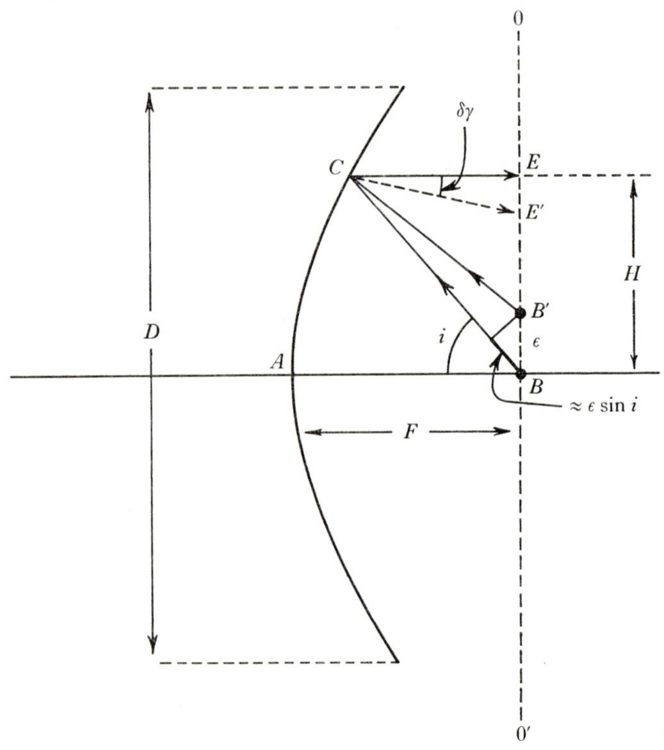

Figure 3.19. Geometrical construction for the calculation of coma distortion. The feed is displaced in the focal plane a distance ϵ from the axis and the direction of maximum response is changed by an angle $\delta\gamma$.

For small values of H/F only the first term of this expression will be important and the path difference $\epsilon \sin i$, which represents a phase difference $2\pi\epsilon \sin i/\lambda = 2\pi\delta\gamma\, H/\lambda$ radians, is linear in H and corresponds to a tilt in the wavefront, i.e. a shift in the beam by an angle $\delta\gamma$ (see (2.57)).

For larger values of H/F, however, the second term, which is of third order in H, cannot be neglected. The maximum error caused by this term $\delta\phi_{\max}$ occurs when, $H = D/2$ i.e.

$$\delta\phi_{\max} = \frac{\pi\delta\gamma D/\lambda}{16(F/D)^2}. \tag{3.29}$$

It is convenient to replace the angle $\delta\gamma$ by an angular displacement $n\theta_{\frac{1}{2}}$, which measures $\delta\gamma$ in terms of the beam width between the half power directions $\theta_{\frac{1}{2}}$. If we put $\theta_{\frac{1}{2}} = 1{\cdot}2\lambda/D = 1{\cdot}2/2a$ which is appropriate for an average paraboloid, we have

$$\delta\phi_{\max} \equiv \frac{0{\cdot}075 n\pi}{(F/D)^2}. \tag{3.30}$$

It would appear to be reasonable to allow a maximum error of π radians since this, with the usual tapered grading of a circular aperture, would be expected to reduce the effective area by only a few per cent while not producing a very great change in the beam shape. In fact, calculations have shown that this value of $\delta\phi_{\max}$ does produce coma sidelobe responses that are about 15 db below the main lobe and are therefore stronger than the normal sidelobe of the telescope.[312] We shall use this value of $\delta\phi_{\max}$, however, as a measure of the maximum safe error in Table 3.3.

Table 3.3. *Displacement (in beamwidths, n) of the response of an antenna for a maximum phase error of π radians.*

F/D	1·0	0·5	0·43	0·25
n	13	3·2	2·4	0·8

From this Table it is obvious that the short-focus paraboloids ($F/D \approx 0{\cdot}4$) used in radio astronomy are not suited to the production of images of a significant size.

3.3.3. *The Cassegrain telescope*

In an attempt to overcome the limitations imposed on image formation by the small F/D ratio of the common radiotelescope, designers have introduced the double-mirror optics of the Cassegrain system into radiotelescopes. A Cassegrain reflecting telescope is shown in Figure 3.20. A large parabolic mirror BE and a smaller hyperbolic reflector CO have a common focus at D. The second foci are respectively at infinity and at E. From the geometry of the surfaces we have for the parabola $AB+BC+CD$ = constant, and for the hyperbola $CE-CD$ = constant. Adding these we find $AB+BC+CE$ = constant, which implies that parallel rays arriving from the axial direction are brought to a focus at E. It is convenient to put E close to the surface of the paraboloid. From Figure 3.20 we see that the effective focal length of the telescope has been increased to $f_1 \cdot f_3/f_2$, which is many times greater than f_1. Thus, a long-focus

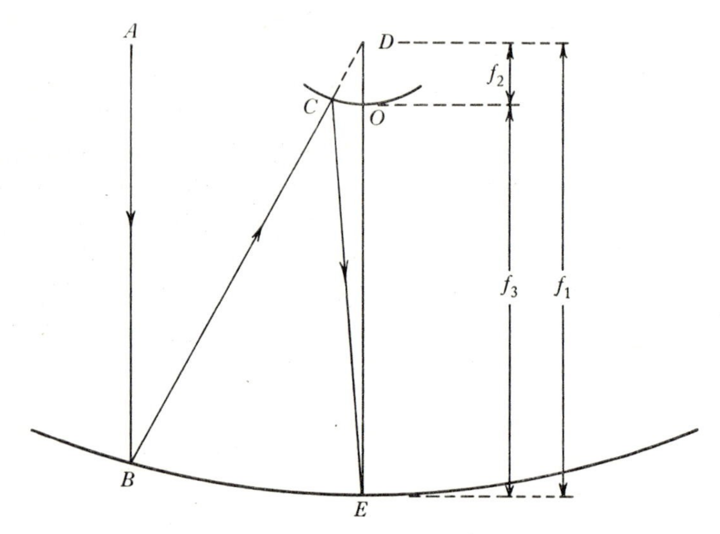

Figure 3.20. The ray path in a Cassegrain telescope consisting of a parabolic primary reflector BE and a hyperbolic secondary reflector CO.

telescope can be produced without any lengthening of the structure in front of the main mirror. Such an increase in focal length might be expected to improve the image-forming capacity of the telescope, and an investigation of the optics shows that this is so. The size (in wavelengths) of the feed antenna at E must, of course, be much greater than is needed for a conventional short-focus antenna, since it must illuminate efficiently the small solid angle occupied by the Cassegrain mirror.

In addition to improved image-forming potentialities the Cassegrain antenna has two other advantages. The first is that the feed antenna may be placed very close to the radio receiver in a mechanically stable and easily accessible location near the apex of the main reflector. This is very important for low-noise receivers such as masers.

The second advantage is that stray pickup of radiation by the feed antenna comes from the direction of the sky, which is relatively cold at short wavelengths, instead of from the relatively hot ground. Hence the amount of thermal noise introduced into the system is kept at a low level. With a Cassegrain telescope, as with all double-reflectors, the surface accuracy of the mirrors must be better than with a single-reflection instrument.

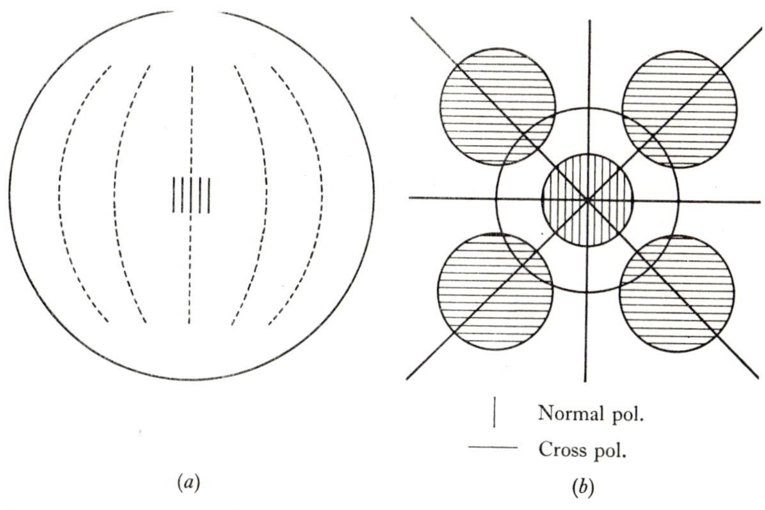

Figure 3.21. (*a*) Barrel distortion of the image of a small vertical grid, produced by a short-focus paraboloidal mirror, illustrates the formation of cross-polarized components in the diagonal quadrants of a reflector. (*b*) The cross-polarized responses cluster diagonally around the maximum response of the telescope. —, Cross-polarized; |, normal.

3.3.4. *Cross polarization*

In the analysis of the performance of a paraboloidal reflector we have assumed that the polarization of the fields at the aperture of the telescope is the same as that of the feed antenna. In fact, this is not quite true. A short-focus paraboloid produces an image which suffers from so-called barrel distortion (Figure 3.21 *a*). This figure illustrates the polarization of waves from different parts of a short-focus reflector. It is apparent that at different parts of the aperture there are horizontally polarized field components. These will produce horizontally ('cross') polarized components in certain directions as shown in Figure 3.21 *b*.[202] Cross-polarization distortion is worst for reflectors which have the smallest F/D ratio.

3.3.5. *Multi-frequency feeds*

One of the great advantages of a parabolic dish reflector is that it can be illuminated by one simple feed antenna. A parabolic antenna is very useful as a multi-frequency device, therefore, since we can change the frequency at which it is used by a simple change of feed antenna. Often simultaneous observations at different frequencies, or else spectral measurements embracing a wide range of wavelengths, are required. For such purposes periodic structures have been built in which the polar diagram does not change significantly with frequency.

The theory of such structures has been developed in recent years.[313–15] The log-periodic, the planar spiral and the conical spiral are the best known of these devices. The first and third have the disadvantage that their centre of radiation changes with frequency and this limits their usefulness as a feed antenna for a paraboloidal reflector.

3.4. Steering the antenna

Fully steerable radiotelescopes are placed on either an equatorial or an altitude-azimuth type mounting. Transit telescopes have limited steerability, usually restricted to the meridian plane.

3.4.1. *The equatorial mount* is the more convenient. One axis is parallel and the other at right angles to the Earth's axis. Rotation around the first axis, at constant angular velocity, counteracts the Earth's rotation and enables the telescope to be pointed in one direction in space. In the other coordinate the telescope can be directed to points in the sky which have different angles of declination (Plate 2, facing p. 96).

The range over which this angle can be varied differs for different telescopes. In general, a design in which a large radiotelescope can be directed anywhere between the celestial pole and the opposite horizon is a costly one.[316] Figure 3.22 shows sketches of two typical radiotelescopes.

In (3.22*a*) and Plate 2 (facing p. 96) a long polar axis is used and there is a great distance between the polar and the declination axes. In this design there is room for large gears to allow accurate driving. There is full counterbalance of all moving parts. The supporting structure is large and expensive. In 3.22*b* the range of angular variation is small in both coordinates, short axes are used and these are placed as close as possible to the centre of mass of the

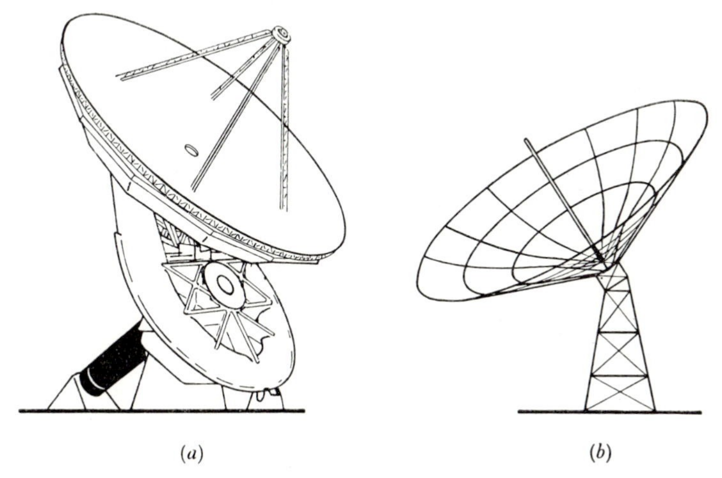

Figure 3.22. Equatorially mounted radio telescopes. (*a*) An accurate and costly telescope capable of observing the whole sky. (*b*) A light and cheap telescope restricted in the range and accuracy of its pointing.

reflector (inside the reflector structure).[317] Counterweights are small and this reduces the necessary strength and weight of the antenna structure. This in turn reduces the fixed supporting structure, which may be small, light and inexpensive. This type of antenna would be used typically for solar observations. (See Plate 3, facing p. 96.)

3.4.2. *The altitude-azimuth mount* has one or two convenient features for astronomical work and a number of inconvenient ones. Very large radiotelescopes, however, are practically all of this type (see Plate 4, facing p. 96), because the vertical and horizontal axes allow much easier and cheaper solutions of the various mechanical problems in the design.

Two examples of this are shown in the sketches of Figure 3.23. Both have a circular horizontal track on which the antenna can rotate in azimuth, but in one case the circular track is larger than the dish and in the other it is smaller.[318–20] The most useful characteristic of the alt-azimuth telescope is that gravitational deflections vary only with changes in altitude angle of the telescope

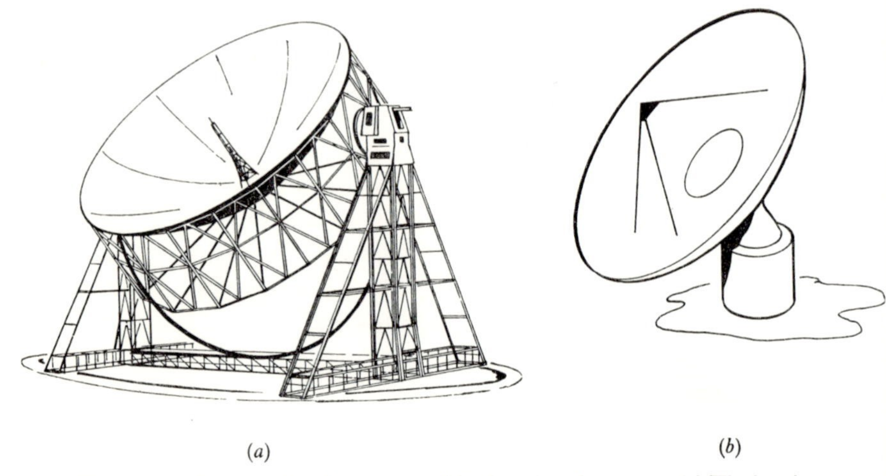

(*a*) (*b*)

Figure 3.23. Large radiotelescopes on altitude-azimuth mounts. (*a*) The bearings for the altitude axis are widely separated and the azimuthal movement is by means of carriages on a large circular rail track. (*b*) The altitude axis and the circular rail track are reduced in size and the range of altitude variations is reduced compared with (*a*).

and are independent of changes in azimuth angle. This makes corrections for flexure very much easier than for an equatorial one where the flexure is a function of both degrees of freedom. The same advantage holds, too, for atmospheric refraction, which is a function of altitude and requires a correction only in this angle.

To these important advantages are opposed several undesirable features. These are the effects of the non-uniform axial motions required to keep the telescope directed towards some point in space. Instead of one rotation at constant angular velocity, as is required with an equatorial telescope, we have two simultaneous angular rotations with the rate of rotation on each axis dependent on the position angles of the telescope on both axes. The rate of rotation in azimuth, for example, approaches infinity as a star passes through the zenith. A computer is needed to keep the antenna in the required direction and to control the varying rate of rotation of the telescope about both axes.

In order to convert angular positions from one set of coordinates to the other, we may calculate the positions by means of certain well-known geometrical relations (Appendix 4). Alternatively, we may build a model of the celestial sphere, rotate it at the celestial rate and have some device which computes the altitude and azimuth angles corresponding to some particular *RA* and declination and passes the information to a servo mechanism which directs the antenna. In a refined form of this device the celestial sphere is replaced by an equatorial telescope which is made to point to some predetermined part of the sky in the normal manner of equatorial telescopes.[(321)] A servo mechanism directs the alt-azimuth radiotelescope to point in the same direction as the equatorial telescope.

A further difficulty with the alt-azimuth telescope is that the field rotates non-uniformly with respect to objects in space during the Earth's daily rotation and the feed antenna at the focus of the telescope must be rotated if observational errors caused by rotation of the plane of polarization of incoming waves are to be avoided. This, however, has some advantages in polarization measurements in allowing the observer to sort out various instrumental effects.

3.4.3. *Transit telescopes*

If we remove the azimuthal movement of an alt-azimuth telescope, the cost of the telescope is considerably reduced. The removal of full steerability from a radiotelescope has three effects:

(*a*) A particular region can be seen only once or twice during twenty-four hours. If the presence of the Sun disturbs the observations, then these observations of some particular region can be made only during certain months of the year.

(*b*) The antenna cannot be used for observation of eclipses of the Sun, lunar occultation of radio sources, disturbances on the Sun or planets, or flare stars, or any other phenomena which cannot be measured at meridian transit.

(*c*) The sensitivity of a transit telescope cannot be increased by following a source over many hours. The most that can be done is to follow the source for an angular distance equivalent to a few beam-widths by moving the feed in the focal plane and by repeating observations each day for many days.

Most, if not all, of the transit radiotelescopes built at the time of writing are of the meridian transit type—i.e. they have a horizontal east–west axis which allows objects to be observed as they pass the meridian. An alternative type of transit instrument would have a fixed elevation angle and be variable in azimuth.

CHAPTER 4

OTHER TYPES OF FILLED-APERTURE ANTENNAS

The steerable paraboloidal reflector described in Chapter 3 is the simplest type of radiotelescope, since all that is required is a passive reflecting surface and an elementary type of feed antenna. The antennas to be described now are electrically more complicated but the mechanical problems are simplified, as these antennas, unlike the steerable parabolic dish, are supported at many points and their weight and cost are approximately proportional to their area rather than to the 3/2 power of the area.

4.1. Arrays

Arrays of antenna elements are used in radio astronomy in many different ways. Most of these arrays fall into the category of unfilled-aperture antennas (Chapter 6) but some, particularly in the case of line apertures, are filled. An array consists of discrete elements the outputs from which are separate, and these outputs can be combined in various ways to steer the response in different directions or even to form an image (many simultaneous beams). Thus, the array can be spread over a larger area than would be possible with any mechanically steerable structure. The individual elements, of course, may be steered mechanically if necessary, but this is much simpler than steering the whole array mechanically as a single unit. Two-dimensional plane arrays of dipoles are used mainly for metric and decametric wavelengths where the apertures may be large physically but not large in terms of the wavelength. The number of elements becomes impracticably large if such an array is to have a high resolving power in both dimensions.

4.1.1. *The field pattern of an array*

The pattern of a general three-dimensional array can be calculated starting from the three-dimensional version of (2.27):

$$f(l,m) = \iiint_{-\infty}^{\infty} g(x,y,z)\exp\{j2\pi(xl+ym+zn)\}\,dx\,dy\,dz$$
$$(n = \pm\sqrt{[1-l^2-m^2]}). \quad (4.1)$$

The three-dimensional grading can usually be expressed as the product of simple gradings in the three orthogonal directions:

$$g(x,y,z) = g(x,0,0)\,.\,g(0,y,0)\,.\,g(0,0,z). \quad (4.2)$$

The triple integral in (4.1) can then be simplified:

$$f(l,m) = \int g(x,0,0)\exp\{j2\pi xl\}\,dx\,.\int g(0,y,0)\exp\{j2\pi ym\}\,dy$$
$$\times\int g(0,0,z)\exp\{j2\pi zn\}\,dz$$
$$= f_x(l)\,.\,f_y(m)\,.\,f_z(n), \quad (4.3)$$

where, as before, $n = \pm\sqrt{(1-l^2-m^2)}$.

$f_x(l)$, $f_y(m)$ and $f_z(n)$ are the one-dimensional Fourier transforms of the three component gradings. The field pattern of the three-dimensional array is then given by (2.28):

$$F(l,m) = \text{const.}\,F_e(l,m)\,.\,f(l,m). \quad (2.28)$$

The Fourier transform of a uniform grading has already been derived (2.34) and (4.3) allows us to write the field pattern of a uniformly graded three-dimensional antenna. However, in an array there cannot be a uniform current distribution; the distribution of the elements will be a grid of points. We can derive $F(l,m)$ in two ways: either we can replace the integrals of (4.3) by a sum of a finite number of terms, as is done in an elementary treatment of arrays, or else we can produce the required gradings by an inverse process of convolution, as is shown in Figure 4.1. In that figure we see that the required grading $g(x,0,0)$ convolved with a small uniform grading $g_u(Nx)$ is equivalent to a uniform grading $g_u(x)$. The convolution theorem then shows that

$$f_x(l) \, . \, d_x \frac{\sin(\pi d_x l)}{\pi d_x l} = N_x d_x \frac{\sin(\pi N_x d_x l)}{\pi N_x d_x l},$$

for example,
$$f_x(l) = N_x \frac{\sin(\pi N_x d_x l)}{N_x \sin(\pi d_x l)}. \tag{4.4}$$

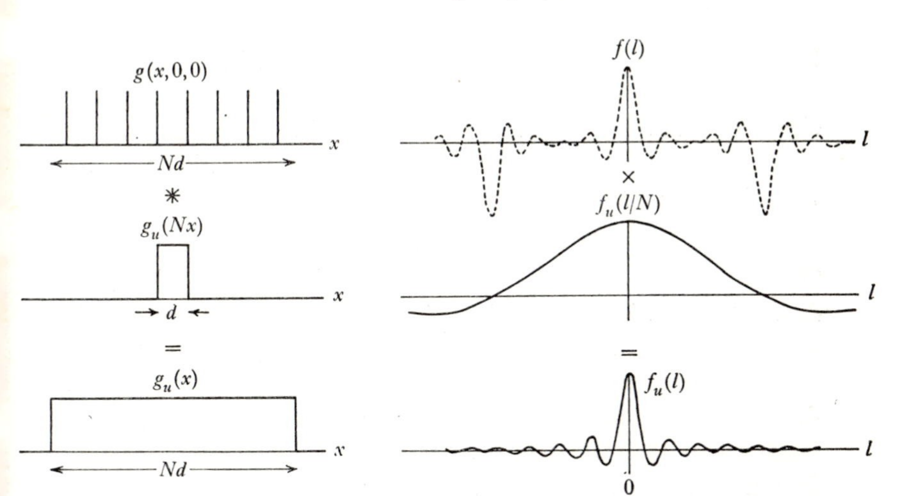

Figure 4.1. A uniform grading $g_u(x)$ is equal to a small uniform grading $g_u(Nx)$ convolved with an array of N elements. Correspondingly the Fourier transform $f_u(l)$ is equal to the product of the other two. We can reverse the process to find $f_x(l)$ for the array from the two transforms of the uniform gradings.

Analogous expressions may be derived for $f_y(m)$ and $f_z(n)$. The field pattern for a uniformly graded three-dimensional array becomes:

$$F(l, m) = \text{const.}\, F_e(l, m) \, . \frac{\sin(N_x U_x)}{N_x \, . \sin U_x} \cdot \frac{\sin(N_y U_y)}{N_y \, . \sin U_y} \cdot \frac{\sin(N_z U_z)}{N_z \, . \sin U_z}, \tag{4.5}$$

where
$$U_x = \pi d_x l,$$
$$U_y = \pi d_y m,$$
$$U_z = \pi d_z n = \pi d_z \sqrt{(1 - l^2 - m^2)}.$$

In practice we cannot make use of a uniformly graded three-dimensional array. In order to have maximum response in, say, the z direction the grading in that direction must have a phase term that changes with z: it cannot be uniform. A common type of three-dimensional antenna consists of a two-dimensional array to which is added its image in a plane reflecting screen. If the distance between array and image is d_z and the phase difference between the object and image is π radians, then, instead of (4.5), we have

$$F(l, m) = \text{const.}\, F_e(l, m) \frac{\sin N_x U_x}{N_x \sin U_x} \cdot \frac{\sin N_y U_y}{N_y \sin U_y} \cdot \sin U_z. \quad (4.6)$$

If $d_z \leqslant 0{\cdot}5$, $\sin U_z$ is maximum in the $n = 1$ direction.

The element field pattern $F_e(l, m)$ depends upon the type of elementary antenna used. The 'thin' half-wave dipole can be calculated exactly and gives a reasonable approximation to half-wave dipoles in general. Its field pattern is

$$F_e(l, m) = \cos(\pi l/2) \,.\, \sqrt{(1 - l^2)}$$

or

$$\cos(\pi m/2) \,.\, \sqrt{(1 - m^2)} \quad (4.7)$$

depending on whether the dipole is lined up parallel to the x-axis or to the y-axis. In such an array, when N is large and d_x and d_y are not greater than one-half wavelength, the field pattern is similar to that of a continuous current sheet. When the element spacings are much larger than one-half wavelength, the field pattern departs from that of the current sheet mainly in the appearance of multiple responses in the field pattern instead of a single response. With equal weighting of all elements of the array and close element spacing the field pattern is approximately that of a uniformly graded aperture. With uniform grading the field pattern sidelobe responses are as great as 0·22 of the main response. In terms of power response this is only about 5 per cent of the main response, which can be tolerated. However, when the array forms part of a correlation system such as we describe in the following chapter, the power in a sidelobe is as high as 22 per cent of that in the principal response of the array, if all elements have the same weighting. Hence, tapering of the array grading must be introduced. We can replace the array grading by that of a continuous sheet and introduce the tapered grading into this.

A very useful family of gradings whose transforms can be calculated exactly is

$$\left.\begin{aligned} g(x) &= (1 - a) + a \,.\, \cos^2(\pi x/L) \\ &= (1 - a/2) + a/2 \,.\, \cos(2\pi x/L). \end{aligned}\right\} \quad (4.8)$$

The transform $f(l)$ of $g(x)$ is easily obtained if we put

$$\cos(2\pi x/L) \equiv \tfrac{1}{2}\exp(j2\pi x/L)+\tfrac{1}{2}\exp(-j2\pi x/L),$$

then
$$f(l) = L(1-a/2)\frac{\sin U}{U}+L.\frac{a}{4}\left(\frac{\sin U'}{U'}+\frac{\sin U''}{U''}\right), \tag{4.9}$$

where $U = \pi Ll, \quad U' = \pi Ll+\pi, \quad U'' = \pi Ll-\pi.$

$F(l)$ is plotted for the grading of (4.8), with $a = 0{\cdot}8$, in Figure 4.2. With this value of a, the sidelobe level is low.

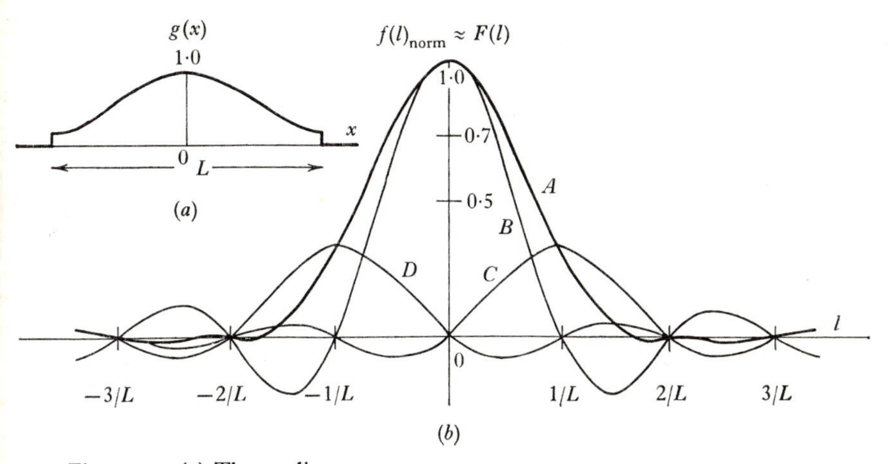

Figure 4.2. (a) The grading

$$g(x) = (1-a)+a\cos^2\left(\frac{\pi x}{L}\right).$$

(b) The Fourier transform A corresponding to this grading is equal to the sum of three gradings B, C, D, each of which has the form $\sin U/U$. The maxima are displaced by an amount $l = 1/L$. The transform A has been normalized to 1·0 at the maximum and hence represents the field pattern of long antennas ($L \gg 1$) with the grading $g(x)$. This relation provides an easy means of removing the effects of sidelobes from records obtained with a $\sin U/U$ antenna characteristic.

To change from the continuous grading of (4.8) to a discontinuous grading representing an array of discrete elements we use the convolution theorem. We proceed in the same way as we did for the uniform-amplitude array shown in Figure 4.1, but it is more illuminating to consider the required grading (Figure 4.3 *c*) as the product of a continuous grading (4.3 *a*) and an infinite array (4.3 *b*). Then the required field pattern (4.3 *f*) is proportional to the convolution of the two patterns (4.3 *d*) and (4.3 *e*). In contrast with

Figure 4.1 we have to deal with a convolution instead of the much simpler multiplication, but both functions have simple forms. If the interval between the elements of the array is 0·5 wavelength, then the separation of the maxima in Figure 4.3*f* is π radians (2.0 in the coordinate l) and only one beam is normally present in the hemisphere of observation. If the separation between elements is d wavelengths, then there will be $2d$ maxima ('grating' beams) in the hemisphere of observation ($l = -1$ to $l = +1$).

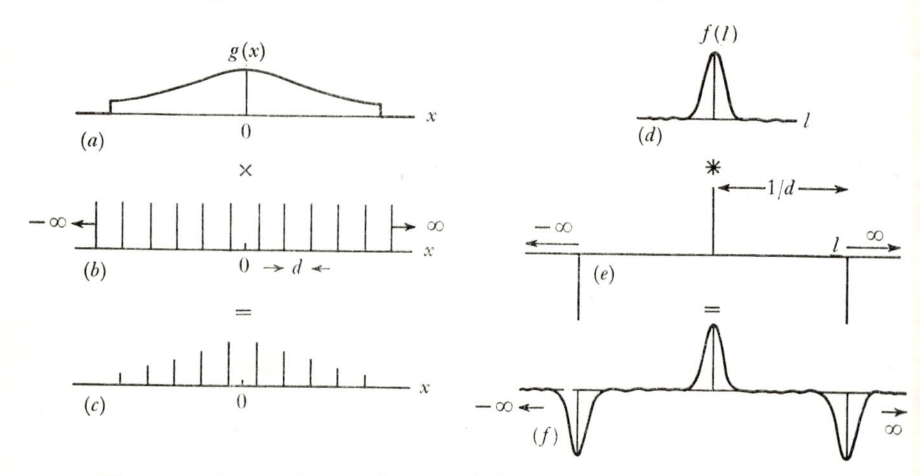

Figure 4.3. A tapered array of separated elements (a tapered grating), (*c*) whose Fourier transform (*f*) can be calculated from the known transforms (*d*) and (*e*) of a tapered continuous grading (*a*) and an infinite grating (*b*). The field pattern of the tapered grating equals the transform (*f*) multiplied by the element field pattern.

4.1.2. *Steering an array*

It can easily be seen that a progressive shift ϕ radians in the phase of the current in adjacent elements of a regular array (Figure 4.4) will turn the response of the array through an angle $(\pi/2-\alpha)$ given by

$$d \, . \cos\alpha = d \, . \, l = \phi/2\pi. \qquad (4.10)$$

The path difference for adjacent elements is then compensated by the phase difference ϕ. This was put more generally in (2.57), which shows that a change from a direction l to a direction $(l+s)$ corresponds to a change in the grading from $g(x)$ to $g(x) \, . \exp(j2\pi sx)$. An array differs, however, from a continuous aperture inasmuch as it has multiple responses, even though, when the element spacing is

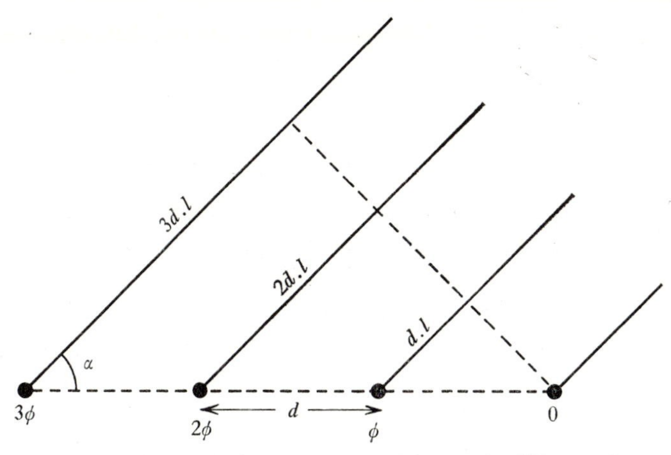

Figure 4.4. The path differences between rays arriving at the different elements of a grating from a direction l may be compensated (at one frequency) by a progressive phase change ϕ along the antenna. The antenna response is thus steered in the direction l.

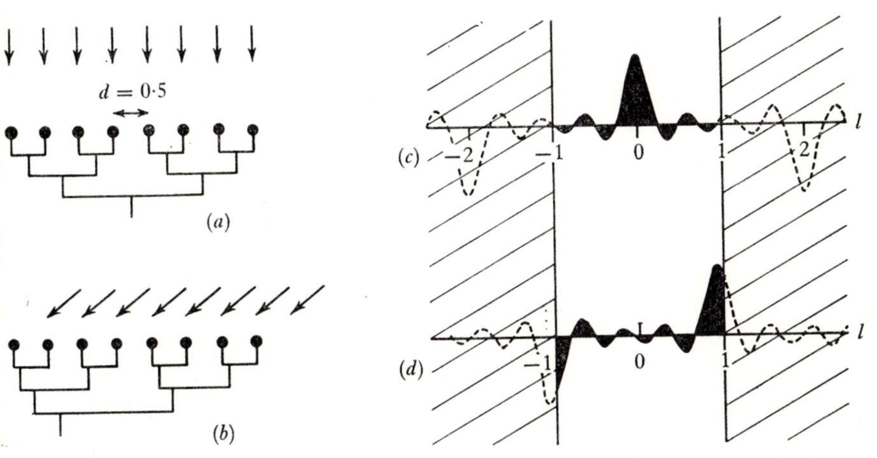

Figure 4.5. The field pattern (*c*) of an untapered grating (*a*) and the modified field pattern (*d*) when there is a progressive phase shift along the grating. The blackened sections of (*c*) and (*d*) are those parts that are inside the horizon.

small, only one of these responses may occur in the hemisphere in which observations are made (see Figure 4.3). In Figure 4.5*d* the array has been phased so that, instead of the maximum occurring at $l = 0$, as in Figure 4.5*c*, it occurs near $l = +1$. Sidelobes from the normally invisible pattern are now appearing at the horizon near $l = -1$.

For an array of closely spaced elements the difficulties involved in steering the response lie (i) in the large number of elements to be phased and (ii) in the electrical coupling between adjacent elements. This coupling means that the current in any one element will induce an e.m.f. in its neighbours; as the phase between neighbouring elements is varied for the purpose of steering, so will the phase of this induced e.m.f. also vary. Each element r of a regular array has adjacent elements in which the current in $(r-1)$ lags by a phase angle ϕ and in the other $(r+1)$ leads by a phase angle ϕ. The result of these currents in $(r-1)$ and $(r+1)$ on the element r is

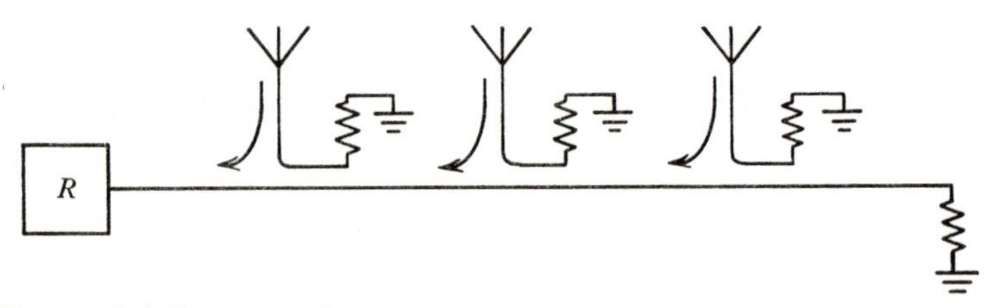

Figure 4.6. A linear array loosely coupled to a terminated transmission line. The array phasing is varied by a change in position of the coupling units.

an induced e.m.f. of constant phase but with an amplitude proportional to cos ϕ. If, for example, the array elements are separated by $\lambda/2$, then, as the array beam is steered from the normal direction through 90°, the induced e.m.f. due to the two surrounding elements varies from a maximum through zero to a maximum of reversed sign. The other elements in the array also contribute to the total induced e.m.f. in a similar way. The impedance of each element, therefore, undergoes a considerable change as the direction of maximum response is changed; in other words, the system cannot remain matched during this process. The mismatch of the elements of the array when viewed from the receiver or transmitter over paths which have different electrical lengths (for phasing purposes) results in different elements presenting different impedances at the receiver or transmitter. This can produce periodic errors in amplitude and phase along the array unless special steps are taken to prevent them. One way to overcome these undesirable effects is to couple loosely by means of directional couplers all the elements of the antenna to a terminated transmission line, as shown in

Figure 4.6.[401] It is arranged that each antenna will act as a high resistance shunt across the line. The directional couplers can be moved along the transmission line so as to steer the antenna response. The weighting of each antenna element can be varied by changing the degree of coupling to the transmission line. Other means we can adopt to steer the antenna response without introducing periodic phase and amplitude errors include the use of isolating devices (electronic amplifiers, ferrite 'isolators') on each antenna element or, alternatively, of phase-rotating devices that do not require varying lengths of transmission line between the receiver and each antenna element (for example, rotatable circularly polarized antennas), or compensating circuits.[402–4]

4.2. Cylindrical paraboloids

A linear array of dipoles is often used as the line feed of a cylindrical paraboloid reflector (Figure 1.2).[405] The antenna beam can be steered in one dimension by rotating the reflector about the cylinder axis. It may be steered electrically in the other dimension in one of the ways we have already described.[401–4] It is possible to use both mechanical and electrical steering and so have the antenna response fully steerable. This has been done in an antenna situated in India near the Equator.[406] The antenna is placed on a N.–S. slope so that its long axis lies parallel to the earth's axis. Changes required to follow a celestial source are made mechanically, while changes in declination are made electrically.

4.2.1. *Mechanical features*

A cylindrical paraboloid which has a tiltable reflector is obviously cheaper per unit area when its two dimensions are greatly different. The cost is directly proportional to the length of the cylinder but increases roughly as the square of the width.

The economy resulting from the use of long thin parabolic cylinders has made these antennas very suitable as component parts in interferometers and crosses, where their fan-beam response can be employed most usefully. A very successful form of construction was devised at Cambridge;[407] in it, parabolic supports were placed at intervals in a line and wires were stretched from one end of the antenna to the other across these supports. The stretched wires formed the reflecting surface. Such a reflecting surface is most useful at metre wavelengths where the wires need not be very closely spaced (see section 3.2.4). The line feed itself can then often be made from stretched wires. The stretched-wire construction makes the reflector effective only for the component of the incoming radiation that is polarized in a direction parallel to the wires. With polarization measurements becoming increasingly important in radio astronomy such reflecting surfaces are becoming less popular.

4.2.2. *Electrical performance*

The long cylindrical paraboloid has characteristics that differ considerably from those of the paraboloid of revolution described in Chapter 3. One of the main differences, of course, is that the feed is much more complicated, has higher ohmic losses, and is less flexible for frequency and polarization changes than the parabolic dish. Balanced against this is the possibility of obtaining multiple responses or producing a one-dimensional image by sectionalizing the feed.

The field pattern of a cylindrical paraboloid is that of a rectangular aperture in which the grading in the direction of the long axis of the cylinder (the x-direction) is independent of the grading at right angles to this (the y-direction)—i.e.

$$g(x,y) = g(x,0) \,.\, g(0,y). \tag{4.11}$$

The Fourier transform of this grading (compare with (4.2) and (4.3) for the three-dimensional case) is

$$\begin{aligned} f(l,m) &= \int_{-L/2}^{+L/2} g(x,0) \exp\{j2\pi xl\}\, dx \,.\, \int_{-D/2}^{+D/2} g(0,y) \exp\{j2\pi ym\}\, dy \\ &= f_x(l) \,.\, f_y(m). \end{aligned} \tag{4.12}$$

The grading $g(x,0)$ is substantially that of the line feed of the cylindrical paraboloid and can be treated as a one-dimensional array, as in section 4.1. The grading in the other direction $g(0,y)$ is determined by the field pattern of the feed in the plane $x = 0$, but is modified by a distance factor (Figure 4.7)

$$\left(\frac{1+\cos\theta}{2}\right)^{-\frac{1}{2}} \tag{4.13}$$

which is proportional to $r^{\frac{1}{2}}$ and is derived in a similar way to (3.24). The distance factor is seen to be less important here than in the corresponding problem for a paraboloid of revolution since it enters only as the square root of r.

A uniform grading with no spillover will, for instance, result

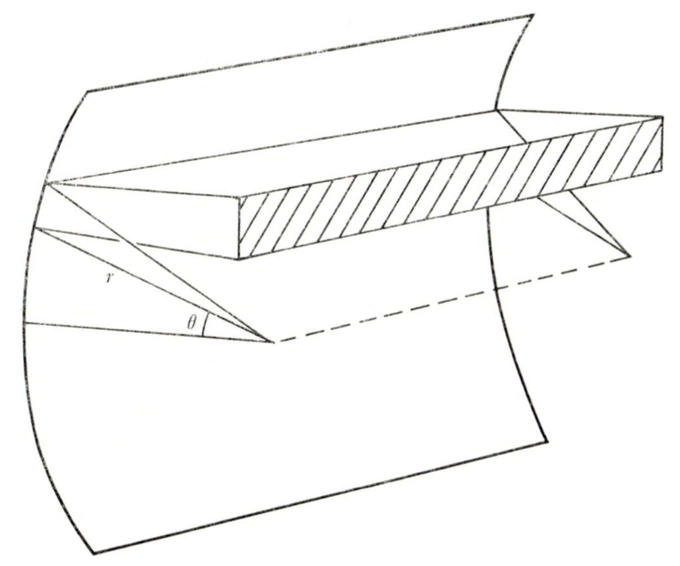

Figure 4.7. Geometry of a cylindrical paraboloid used for calculating the effect of the variation in distance from focus to different parts of the reflector on the grading of the aperture.

from a feed whose field pattern in the plane at right angles to the line is given by

$$F_F(0, m) = F_F(0, \sin\theta) = \begin{cases} \left(\dfrac{2}{1+\cos\theta}\right)^{\frac{1}{2}} & \text{for} \quad |\theta| \leqslant \theta_{\max}, \\ 0 & \text{for} \quad |\theta| \geqslant \theta_{\max}. \end{cases} \tag{4.14}$$

The field pattern of the feed is proportional to the product of the element pattern and the Fourier transform of the grading of the feed (2.28). We can get a rough impression of how the feed should be constructed in order to give uniform illumination by ignoring the relatively slow dependence on the element pattern and the distance factor. The problem is then to find the grading across the feed line

$g_F(0, y)$ whose Fourier transform is uniform for all directions $|m| = |\sin\theta| \leqslant m_{\max}$ and zero for $|m| > m_{\max}$. The grading itself is then the inverse Fourier transform of the desired pattern (see (2.23), (2.24) and Appendix 2)—i.e. in this approximation

$$f_y(m) \propto F_F(0, m) = \begin{cases} 1 & \text{for} \quad |m| \leqslant \sin\theta_{\max}, \\ 0 & \text{for} \quad |m| > \sin\theta_{\max}, \end{cases}$$

$$g_F(0, y) = \text{const.} \int_{-\sin\theta_{\max}}^{+\sin\theta_{\max}} \exp\{-j2\pi y m\}\, dm$$

$$= \frac{\sin(2\pi \sin\theta_{\max} \cdot y)}{2\pi \sin\theta_{\max} \cdot y}. \tag{4.15}$$

The grading has, as usual, been normalized to unity at maximum. It is not possible to build such a feed, of course, since it would extend to infinity. It can, however, be approximated by a feed of finite width, but the approximation will cause the illumination to be not quite uniform and will make the field pattern of the feed extend outside $|m| = \sin\theta_{\max}$—i.e. introduce some spillover.

It is easier to design a feed for a cylindrical paraboloid reflector than for a paraboloid of revolution, because the grading components in the two directions can be controlled independently. It is also easier to achieve a low spillover. There are, on the other hand, several practical disadvantages. One of these is the difficulty in supporting the line feed. There is always some sagging between supports in all such line feeds and this sag changes as the antenna is tilted in different directions. The sag is periodic and will produce grating-type sidelobe responses. Figure 4.8*a* shows a line feed with a cosinusoidal variation in its distance from the focal line. The amplitude of this variation is Δz wavelengths, which produces a maximum phase variation $\beta = \pm 2\pi\Delta z$ radians along the x-axis of the aperture. Then the grading $g(x, 0)$ becomes

$$g(x, 0) = \exp\{j\beta \cos(2\pi x/d)\} \tag{4.16}$$

$$\approx 1 + j\beta \cos(2\pi x/d) \quad \text{when} \quad \beta \ll 1.$$

Since the phase modulation β must always be small in any useful telescope, equation 4.16 shows that we can express the grading approximately as the sum of two gradings with constant phase. One is a uniform grading of unit amplitude and the other is the 'error' grading in which the phase differs from the first by $\pi/2$ and the amplitude is $\beta \cos(2\pi x/d)$. In transmission, the uniform grading component gives rise to a (normalized) far field $F(l, 0)$:

$$F(l, 0) = \frac{\sin(\pi Nd.l)}{\pi Nd.l}, \tag{4.17}$$

Nd (wavelengths) is the total length of the line feed. The cosine

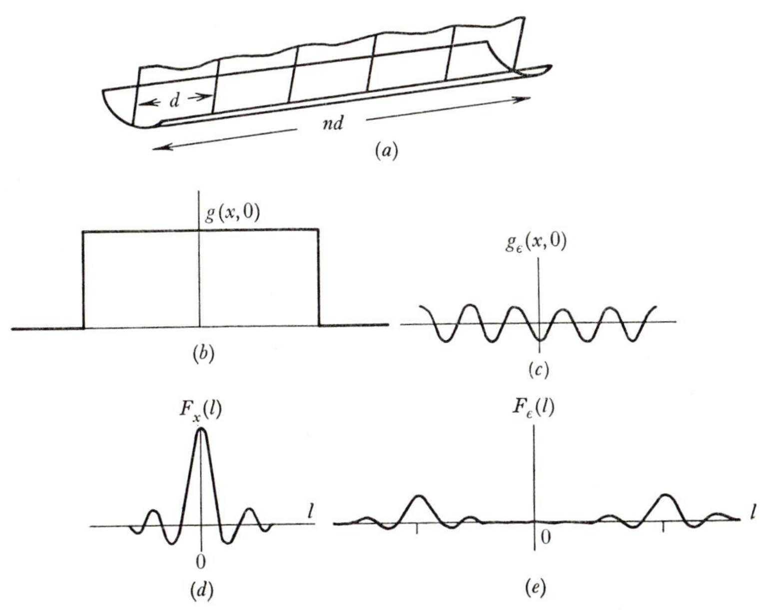

Figure 4.8. (*a*) The sag between supports of a line feed can be represented by an error modulation (*c*) of the ideal grading (*b*). The effect on the field pattern is to produce error sidelobes (*e*) on either side of the main response (*d*).

modulated grading component $g_\epsilon(x, 0)$ causes an error contribution $F_\epsilon(l, 0)$:

$$F_\epsilon(l, 0) = j\beta \int_{-Nd/2}^{Nd/2} \cos(2\pi x/d) . \exp(2\pi xl)\, dl$$

$$= \frac{j\beta}{2}\left[\frac{\sin\{\pi N(d.l-1)\}}{\pi N(d.l-1)} + \frac{\sin\{\pi N(d.l+1)\}}{\pi N(d.l+1)}\right]. \tag{4.18}$$

This particular error will, therefore, give rise to two field pattern sidelobes which have the same shape as the main response and which are separated from it in the coordinate l by $\Delta l = \pm 1/d$. A relatively small periodic error, for example $\Delta z = 0{\cdot}1$ wavelengths, will produce field pattern sidelobes whose amplitude $\beta/2 = 2\pi . 0{\cdot}1/2$ is more than 30 per cent of the main lobe. The corresponding sidelobes in the power pattern have, of course, an amplitude equal to the square of this value, or about 10 per cent.

Another disadvantage of the cylindrical paraboloid compared with the paraboloid of revolution lies in the greater proportion of the aperture blocked by the feed. The broad negative response in the field pattern has an amplitude relative to the main response which, for uniform grading, is equal to the ratio of the blocked area of the aperture to the unblocked area (see section 3.1 and eqns. 3.8 and 3.9). For example, if one-tenth of the aperture is blocked by the feed, then the negative sidelobe in the field pattern has an amplitude of about 11 per cent of the main response. The maximum effective area A_{max} is at the same time reduced to 0·81 of the value it would have in the absence of blocking. The effects in practice will be even more serious, since the grading will usually be non-uniform and the feed be situated above the area of maximum current. The paraboloid of revolution has a point focus and the feed does not, as a rule, block nearly such a large area as it does with cylindrical paraboloids of moderate size.

In correlation telescopes where the power response of the combination is proportional to the product of the field patterns of the component antennas the extended field pattern sidelobe may have serious disadvantages. One way of overcoming this is by placing the feed outside the aperture.[408–9] To do so it is necessary to use as a reflector part of a parabola which does not include the apex, as shown in Figure 4.9. With such a reflector the aperture can be kept free from obstructions. The arrangement shown in the figure has the great advantage that all parts of the reflecting surface and also the feed may be brought very close to the ground during construction and for adjustment. The main disadvantage is that the cost is higher than for conventional cylindrical paraboloids. In addition, a more highly directional feed is required. However, as it is out of the aperture, there is no great difficulty in constructing a large feed.

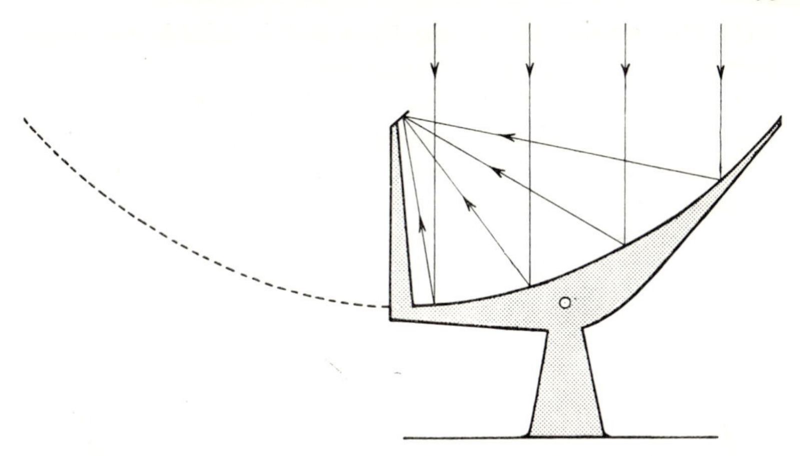

Figure 4.9. A cylindrical paraboloid with an off-axis reflector.

4.3. Special types of parabolic-reflector antennas

4.3.1. *The Kraus type fixed parabolic reflector* with a tilting plate to deflect the response of the antenna to different parts of the sky in the meridian plane has been used in two different forms. These are shown in Figure 4.10 and Plates 5 and 6 (facing p. 96). In form *a*, at Ohio, the ground-plane is a flat highly conducting surface and the feed *A* need have only one-quarter of the aperture (in the vertical plane) that would be necessary if the antenna were arranged as in 4.10*b*, where the response of the feed antenna is nearly zero in all directions that would strike the ground.[410–12]

The fixed antenna *B* is a section of a paraboloid of revolution and *C* is a plane reflector. The antenna has a relatively large F/D ratio, which reduces coma distortion and makes it possible to follow a source for some period of time by moving the feed. An antenna at Nançay is of the type shown in Figure 4.10*b*. The fixed reflector *B*, however, has a parabolic cross-section in the vertical plane and a circular cross-section in the horizontal plane. This allows the feed antenna to be moved along a track to follow a source for about one hour.

Some of the features of the Kraus type antenna are:

(*a*) All active and curved parts of the antenna are stationary, the only moving surfaces are flat, and a flat surface is more easily controlled and monitored than a curved one.

(*b*) The antenna has a long focus and is suitable for image formation, particularly in one dimension.

(*c*) The antenna is well suited to multi-frequency operation.

The features which are less advantageous are:

(*d*) It is primarily a transit instrument, although a source may be followed for an hour or so.

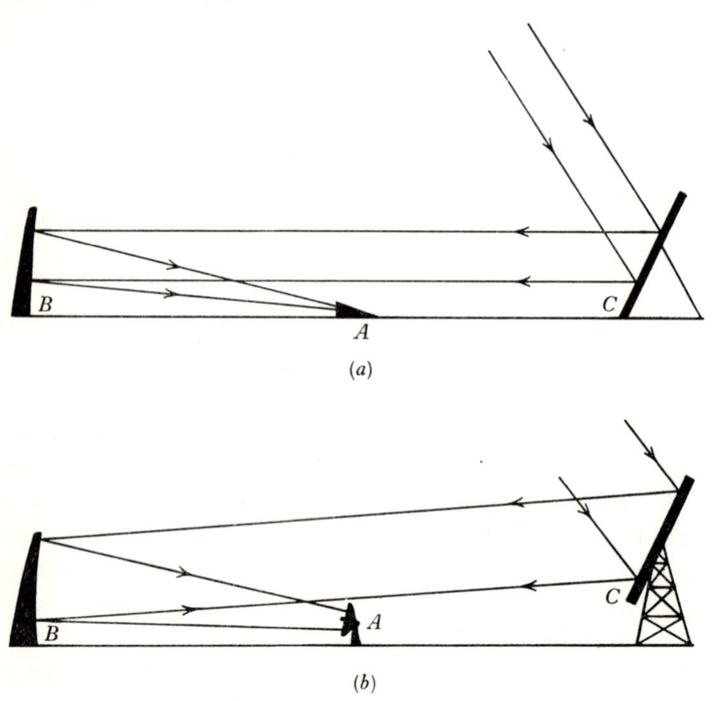

Figure 4.10. Kraus type radiotelescopes. (*a*) The original form at Ohio with the ground forming part of the telescope. (*b*) A modified form used at Nançay with the earth outside the ray paths.

(*e*) It is most economical to make the reflecting surfaces long and narrow—i.e. the response tends to be of the fan-beam rather than the pencil-beam type.

(*f*) The useful range of angles of declination of the antenna is not much greater than 100°.

(*g*) The multiplicity of reflecting surfaces and the proximity of the transmission path to the ground make the problem of spillover rather serious.

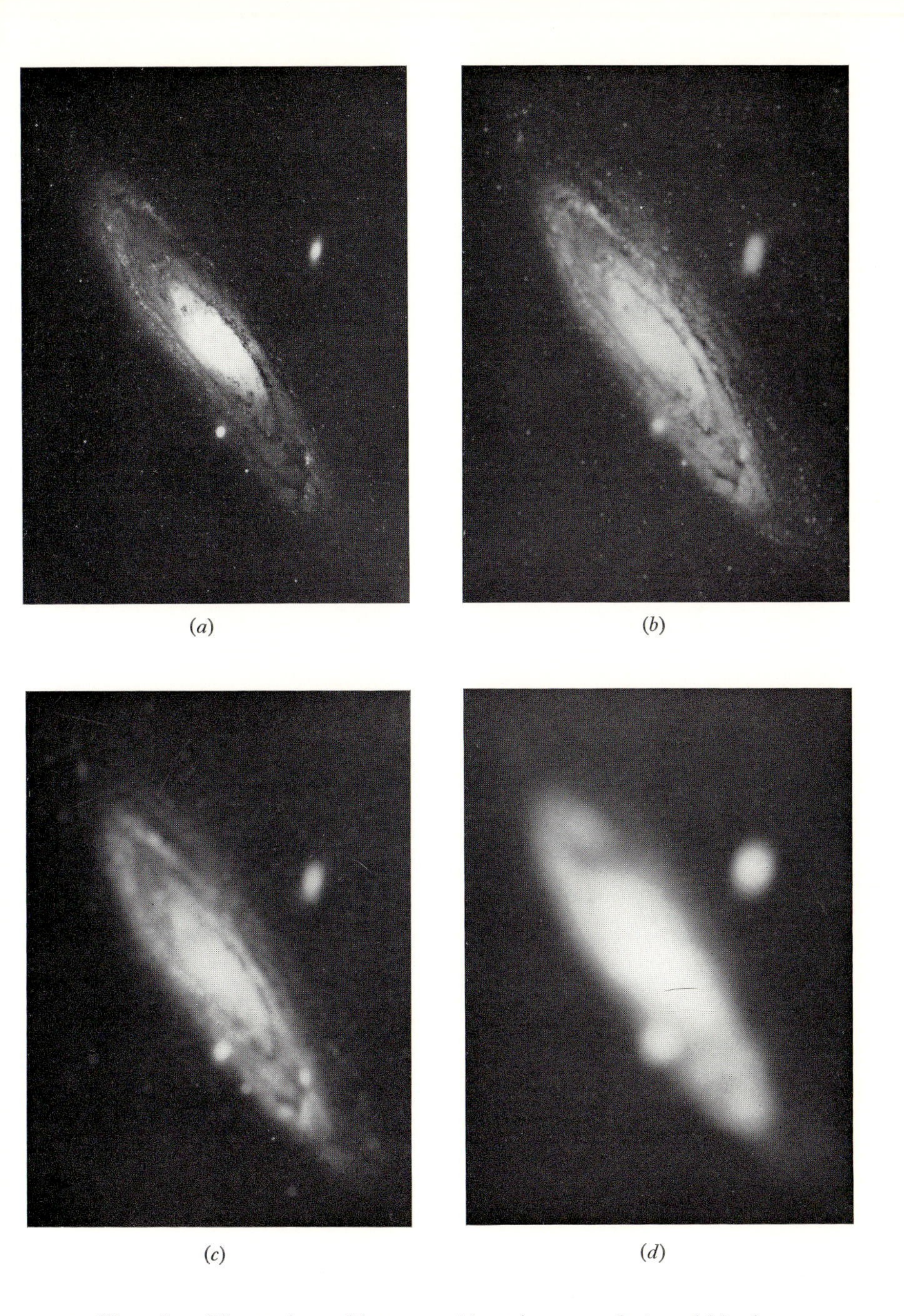

1 The galaxy M31 as it would appear with an image resolution of (*a*) a large optical telescope, (*b*) 1 minute of arc, (*c*) 3 minutes of arc, (*d*) 12 minutes of arc (Sterrewacht, Leiden).

(*facing p.* 96)

2 Equatorially mounted, 43 m diameter parabolic reflector telescope. Greenbank, West Virginia, 1965 (NRAO, Greenbank).

3 Ring telescope: Some of the 96 paraboloids each 12 m aperture arranged in a circle 3 km in diameter. Culgoora, New South Wales, 1967 (Radiophysics Division, CSIRO).

4 64 m diameter parabolic reflector telescope. Parkes, New South Wales, 1963 (Radiophysics Division, CSIRO).

5 Kraus radiotelescope 110 m × 24 m, 1961 (Ohio State—Ohio Wesleyan Radio Observatory).

(*a*)

(*b*)

6 Part of Nançay telescope (Dr J. L. Steinberg, Paris Observatory).

7 The Pulkovo radiotelescope Pulkovo Observatory, Leningrad, 1958 (Pulkovo Observatory).

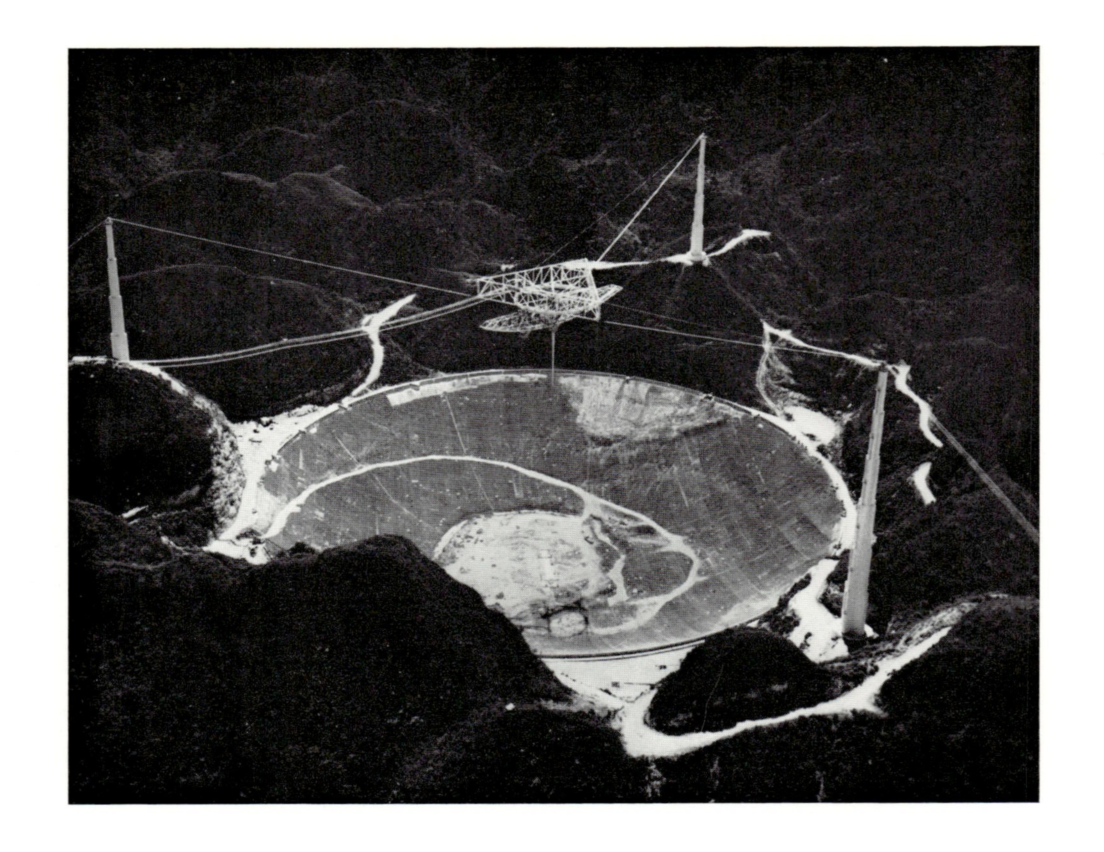

8 300 m diameter spherical reflector telescope. Arecibo, Puerto Rico, 1963 (Cornell University).

9 The first grating radiotelescope Potts Hill, Sydney, 1951, with the second grating 1953. These gratings were used in combination as a rotational synthesis instrument (Radiophysics Division, CSIRO).

10 Compound interferometer, Ottawa, 1958 (A. E. Covington, National Research Council, Canada).

11 (*a*) Westerbork Synthesis Radiotelescope.

11 (*b*) Equatorial Mount of Westerbork 25 m antenna.

(*a*)

(*b*)

12 One Mile cross radiotelescope, Hoskinstown, New South Wales, 1966 (University of Sydney).

13 Grating cross telescope, Medicina, 1966 (University of Bologna).

14 Mobile antenna (60 m × 20 m) of aperture synthesis radiotelescope. Cambridge, 1958 (University of Cambridge).

15 ‘Fleurs’ rotational–synthesis compound grating telescope, Fleurs, Sydney, 1968 (University of Sydney).

4.3.2. *The multiple-plate antenna*, in which a large number of flat surfaces are arranged to form part of a paraboloid of revolution (Plate 7, facing p. 96), was used first at the Pulkovo Observatory.[413] Figure 1.5*b* shows a sketch of this antenna. The antenna response can be steered in the sky by rearranging the individual flat plates as shown in Figure 4.11. In order to change the direction of the response from the horizon (*a*) to the zenith (*c*) the plates must be tilted through 45° in the vertical plane, and the arrangement in the horizontal plane varies between a parabolic shape (*a*) and a circular arc (*c*).

At the same time the angular response changes. In 4.11*a* the response is fan-shaped, with high directivity in azimuth and low directivity in the vertical plane. In the position shown in Figure 4.11*c* the directivity in the two planes has become nearly the same. The sidelobes of the antenna in position (*c*) are higher than in (*a*) and are comparable with those of the ring antenna described in Chapter 6.

The advantages of the Pulkovo antenna are that the focus and the reflecting strip all lie in a horizontal plane and are supported by the ground. Hence it is an economical way in which to construct an antenna of high directivity. More complicated types of multiple-plate antennas lose this fundamental advantage.[414]

Figure 4.12 shows a paraboloid of revolution made up from flat plates. The feed is situated on a tall mast and the flat plates are arranged at ground level. The plates must be arranged in zones with a step between each zone. The height of the step is such as to change the path length of a ray by an integral number of wavelengths with respect to those reflected from other zones. This, of course, makes the antenna frequency-selective and, apart from its great complexity, is its chief drawback.

4.4. Spherical reflectors

A spherical reflector when compared with a parabolic one has two important features:

(*a*) It has the disadvantage that it needs a secondary device to bring parallel rays falling on its surface to a common point.

(*b*) It has the very great practical advantage that it has no principal axis and a stationary spherical mirror can deal equally well with waves from different directions in the sky. The use of a

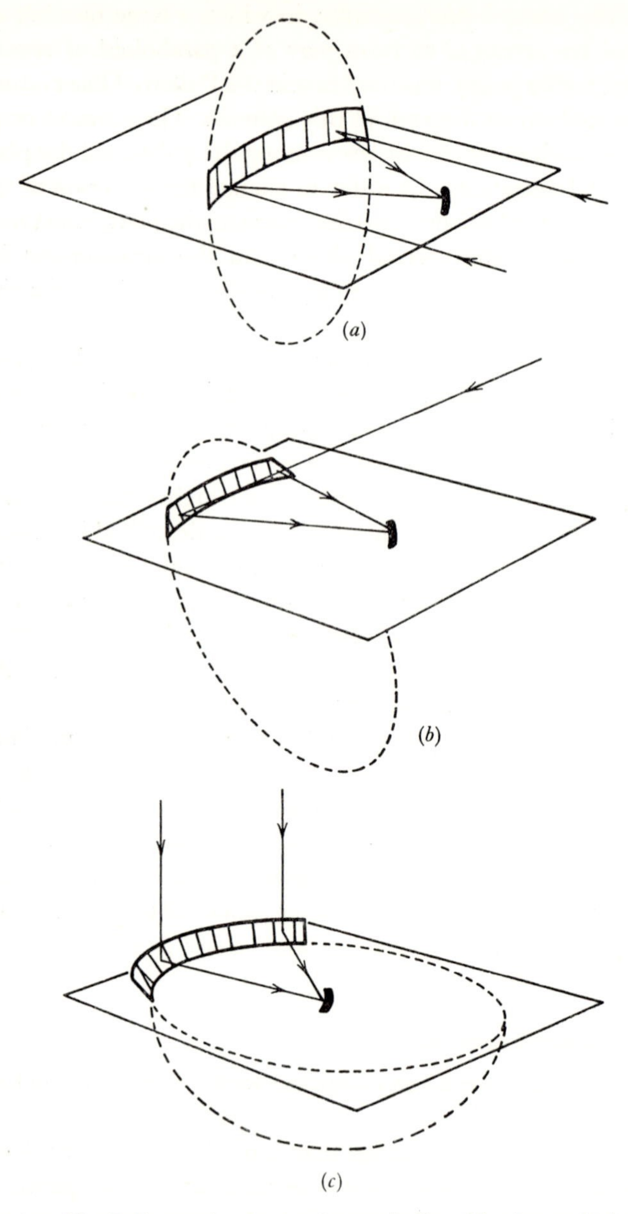

Figure 4.11. The Pulkovo telescope made up of adjustable plates which can be rearranged for reception at different angles of elevation. (*a*), (*b*) and (*c*) show the plates forming part of three different paraboloids.

spherical reflector also opens the way to the use of image-forming devices that would be impossible with a parabolic reflector.

Spherical mirrors have been used for many years in optical telescopes, with either a secondary correcting mirror or a correcting lens. Such telescopes can produce an image of a much greater angular size than is possible with a parabolic reflector. They can be steered mechanically,[415] but for special purposes a fixed hemispherical reflector has been used which can produce an optical image

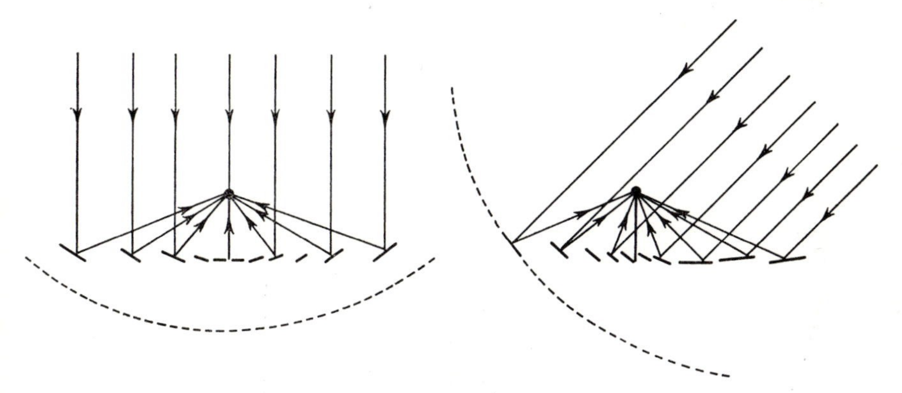

Figure 4.12. A two-dimensional form of the adjustable multi-plate antenna. The feed is placed on a high mast and the plates are arranged along the ground.

of a large part of the sky. It was the fixed-reflector form that was first suggested for use in radio astronomy;[416] a movable secondary mirror or other device brings the rays from a distant source to a focus. To date, no use has been made of the great potential that this type of radiotelescope has for image formation.

4.4.1. *Correcting mirrors*

In order to equalize the lengths of the optical paths from all points in the aperture of a spherical mirror to a single point or focus we may use a secondary reflecting surface.[416–17] In fact, there is considerable freedom in designing such a surface, since there are two parameters (including the position of the focal point) which can be chosen at will. The family of curves representing the secondary reflector can be written in terms of ρ and ϕ, the polar coordinates of the surface with respect to the focus,

$$\left.\begin{aligned} \phi &= 2\theta + 2\arctan f(\theta), \\ \rho &= \tfrac{1}{2}R(\sin\theta - C\sin 2\theta)\,[f(\theta) + 1/f(\theta)], \end{aligned}\right\} \tag{4.19}$$

where $$f(\theta) = (\sin\theta - C\sin 2\theta)/K - 2\cos\theta + C\cos 2\theta. \tag{4.20}$$

ρ, θ, ϕ and C are shown in Figure 4.13. K is an arbitrary constant which can be adjusted to obtain optimum characteristics of the telescope in terms of the position and size of the secondary mirror. In Figure 4.13 one member is shown of the family of curves

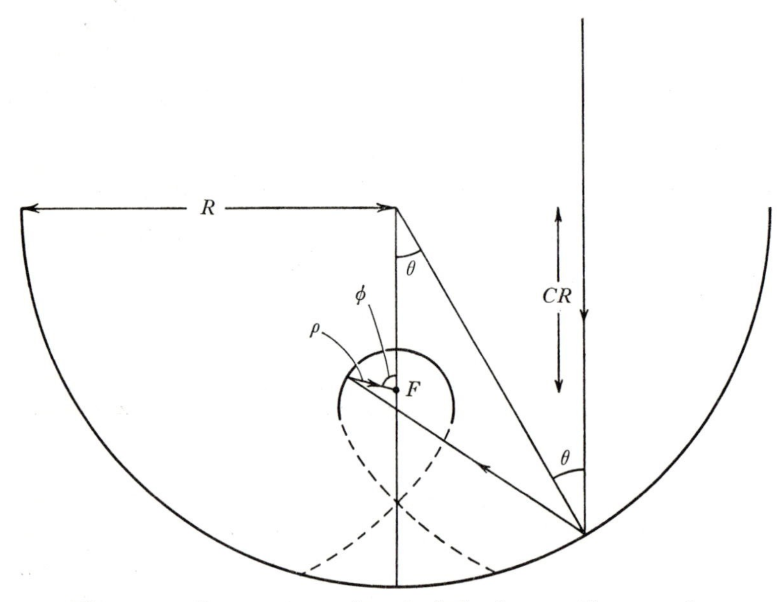

Figure 4.13. Ray geometry of a spherical reflector with a secondary correcting mirror.

required to bring rays to a focus at some selected point F. With different values of K in (4.20) (i.e. with different values of the length of the ray path from aperture to focus) we have different curves. It is obvious that only a section of the curve of Figure 4.13 can be used to form a correcting mirror, since part of the curve forms a closed surface. In Figure 4.14 sufficient of the surface A is used to collect rays over a range of angles $-30° < \theta < +30°$. If the primary reflecting surface is a fixed hemisphere, then the effective width of the aperture of the telescope is about half that of the hemisphere. The correcting mirror may be moved about the centre of the sphere and can accept radiation from all angles within 60° from the zenith.

It should be noted that the correcting mirror blocks a considerable part of the aperture; this is exaggerated in Figure 4.14. We have dealt with the effect of aperture-blocking in section 3.1.

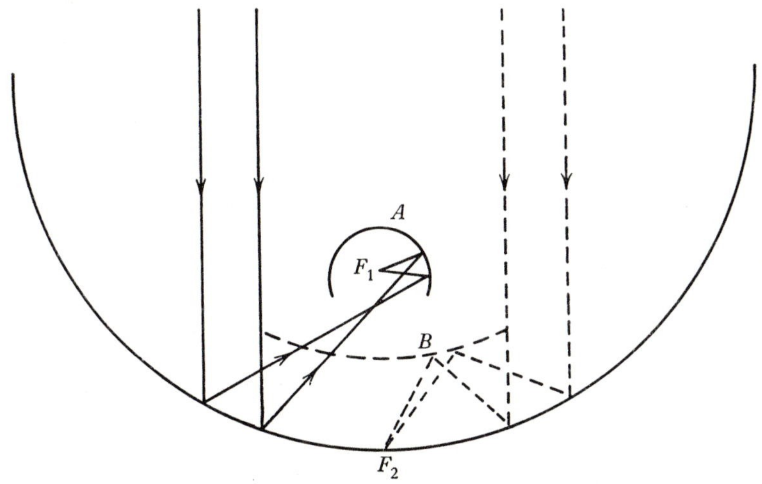

Figure 4.14. Two positions for the correcting mirror of a spherical reflector. *A* brings rays to a focus at F_1 and *B* brings them to a focus F_2 at the surface of the primary mirror.

Figure 4.14 shows another correcting surface *B*. This is one of a family of reflecting surfaces which will bring rays to a focus at a point F_2 near the surface of the hemisphere. It has a number of advantages over the type shown at *A*, since it is very close to the primary mirror and may be supported from it. It is more difficult to feed than the type *A*, since the response of the feed is confined to a narrower range of angles.

4.4.2. *Line feed for a spherical reflector*

Figure 4.14 shows ray paths when θ is about 30°. It is obvious that all these rays pass through the line F_1F_2. Hence it must be possible to design a line feed which will intercept all the energy crossing this line and convey it in the right phase to some common point.[418–19] If we are interested in collecting the energy falling on that part of the hemisphere between $-45° < \theta < +45°$, then the feed must extend from $0{\cdot}5R$ to $0{\cdot}7R$ approximately. For example, if $R = 150$ m, the feed must be 30 m long (Plate 8, facing p. 96). The phasing of adjacent elements along the line feed must be such that the response of the section is directed towards that part of the hemisphere from which energy is arriving. The section closest to F_2 (Figure 4.14) must have a response directed at right angles to the line F_1F_2, and the section closest to F_1 has its maximum response in the direction F_1F_2.

The line feed acts in a somewhat similar way to the correcting lens in a Schmidt camera used in optical astronomy. It is, however, frequency-selective, which places it at a disadvantage when compared with a correcting mirror. Its advantages are its small size and the negligible blocking of the aperture.

4.5. Future developments

A paraboloid of revolution must be steered mechanically and the structure will be subjected to varying gravitational stresses. The normal short-focus paraboloid has, moreover, a very poor image-forming potential. The telescopes described in this chapter have been aimed at overcoming the first disadvantage.

In optical astronomy there has been considerable effort expended in devising optical reflecting systems which have an image-forming capacity greatly superior to that of the paraboloid of revolution. These new optical systems are tending to displace the older paraboloids. It would seem that this same tendency must soon be seen in radio astronomy, if the filled-aperture telescope is to compete with the image-forming radiotelescopes to be described in the following chapters.

CHAPTER 5

SOME MORE THEORY

5.1. Introduction to correlation telescopes

The single antenna ('total-power') radiotelescope absorbs power from the incoming radio waves and makes this power available for measurement by the total-power receiver. The *correlation radiotelescope* consists of two separate antennas connected to a *correlation* (*or multiplying*) *receiver* which is designed to give an output proportional to the average *product of the voltages* from the two component antennas.(501)

Let the open circuit noise voltages $V_1(t)$ and $V_2(t)$ from the antennas contain the components $V_{c1}(t)$ and $V_{c2}(t)$ which are of common origin and therefore correlated. These could, for instance, be due to the radiation from a source which contributes to both antenna outputs. Let the remaining uncorrelated components be $V_{u1}(t)$ and $V_{u2}(t)$. The common origin of the voltage components $V_{c1}(t)$ and $V_{c2}(t)$ is expressed by the equation

$$V_{c2}(t) = \text{const.}\, V_{c1}(t+\tau). \tag{5.1}$$

The constant takes into account any differences in amplitude of the components which may be the result of unequal size or efficiency of the two antennas. The radio (electrical) paths from the source to the receiver via the two antennas are not necessarily equal and this will cause a difference τ (sec) in the time of arrival of the signals.

The time average of the product of the voltages from the two antennas can be written

$$\begin{aligned}\overline{V_1 V_2} &= \overline{(V_{u1}+V_{c1})(V_{u2}+V_{c2})} \\ &= \overline{V_{u1}V_{u2}}+\overline{V_{u1}V_{c2}}+\overline{V_{c1}V_{u2}}+\overline{V_{c1}V_{c2}}.\end{aligned} \tag{5.2}$$

All terms except the last contain the products of uncorrelated voltages and their averages over a certain time interval t will disappear as $t\to\infty$. Hence, if t is chosen sufficiently large we can write:

$$\overline{V_1 V_2} = \overline{V_{c1}V_{c2}}, \tag{5.3}$$

that is, the time average of the voltage product is equal to that of their correlated components. The correlation (multiplying) receiver is designed to measure this average voltage product and the ideal correlation telescope will therefore be sensitive only to signals which are received by both component antennas simultaneously.

The averaging period t will, of course, in practice always be finite and the average product of the uncorrelated voltages will not be quite zero. Equation 5.3 will contain an error term ϵ which represents the non-zero mean of the averaging process. This error will vary in a random way from one t second measurement to the next and is the main reason for the 'noise fluctuations' which limit the sensitivity of the correlation telescope. The sensitivity problems will be dealt with in Chapter 8.

The signal is usually very small compared with the unwanted noise powers in the system. Total-power measurements, then require very stable equipment, because small changes in the level of this unwanted noise can completely mask the changes produced by a radio source passing through the antenna beam. The stability requirements can be considerably relaxed for correlation telescopes, because these are, in principle, insensitive to most of the unwanted noise powers which, generally, contribute only to the uncorrelated voltage components.

The relative immunity of the correlation receiver to irrelevant noise powers opened the way for novel types of radiotelescopes:

(*a*) *Skeleton telescopes*, which are based on the fact that there is a great deal of redundancy in the information gathered by a large-aperture single telescope and that only a fraction of the whole aperture needs to be built, and

(*b*) *Aperture synthesis telescopes*, which make use of a second fact: that only a small part of this minimum required aperture need be present at any one time; the necessary information can be gathered from measurements with smaller antennas if these can be moved so as to make available all spacings and orientations in the desired aperture. This type of telescope is useful only for observation of sources which do not vary significantly in the time it takes to complete the measurements.

Skeleton and aperture synthesis telescopes can be, and have been, operated with total-power receivers, but, as we have already pointed out, the difficulties with stability and other changes in the unwanted noise level are then much greater.

The first correlation receiver to be used in radio astronomy, the so-called 'phase-switching' receiver, measures the average voltage product as the difference between two total-power outputs.[501] Let the two component antennas have the same impedance and be joined together electrically to give a single output. The available power from the combination is then proportional to $\overline{(V_1+V_2)^2}$. If the phase of the signals from the second antenna is changed by π radians, e.g. by inserting an extra half wavelength piece of transmission line into this side of the junction, then the available power will be proportional to $\overline{(V_1-V_2)^2}$. The difference

$$\overline{(V_1+V_2)^2}-\overline{(V_1-V_2)^2} = 4\overline{V_1 V_2} \tag{5.4}$$

is proportional to the desired voltage product. In practice the signal phase may be reversed at a frequency of a few hundred Hertz and and the corresponding modulation of the total power output is measured with a phase-sensitive detector circuit. Other types of correlation receivers, some of which contain circuits for direct multiplication of noise voltages, are now more commonly used than the phase-switching receiver, but the principle of operation of a correlation receiver remains the same.

5.2. The effective area of a correlation telescope

5.2.1. *Definition in terms of the available correlated power*

The effective area of a total power telescope was defined by (2.1)

$$p = AS_m \quad (\mathrm{W\,Hz^{-1}}). \tag{2.1}$$

p is here the available power per unit bandwidth at the antenna output terminals produced by the matched polarized flux density S_m ($\mathrm{W\,m^{-2}\,Hz^{-1}}$.) The equation defines the effective area A($\mathrm{m^2}$) in the particular direction from which the radiation is coming. Let the two component antennas of a correlation telescope have the same polarization and the available powers per unit frequency band due to radiation from a point source be p_1 and p_2 respectively. Then

$$p_1 = A_1 . S_m = \mathrm{const.}_1\, \overline{V_{c1}^2} \quad (\mathrm{W\,Hz^{-1}}),$$

$$p_2 = A_2 . S_m = \mathrm{const.}_2\, \overline{V_{c2}^2}. \tag{5.5}$$

The values of the constants will depend on the antenna impedances and on the frequency bandwidth over which the measurements are made. We define the *available correlated power per unit bandwidth* p_c and the *effective area of the correlation telescope* A_c by the analogous equations

$$p_c = A_c . S_m = \text{const.}\, \overline{V_{c1}V_{c2}} \quad (\text{W Hz}^{-1}). \tag{5.6}$$

In discussions of the correlation telescope it will be assumed that the two component antennas have the same polarization. If this were not so, then the quantity S_m would not refer to the same component of the incoming waves in (5.5 to 5.6).

5.2.2. *Relation between* A_c *and the properties of the component antennas*

The relation between the effective area of the correlation telescope and that of each of the two component antennas can be derived by applying the equation

$$\overline{V_{c1}V_{c2}} = (\overline{V_{c1}^2} . \overline{V_{c2}^2})^{\frac{1}{2}} \cos\psi \tag{5.7}$$

which is easily shown to hold when V_{c1} and V_{c2} are sinusoidally varying voltages which arrive at the multiplying point with a constant phase difference ψ radians. If this phase difference is due only to the difference τ sec in the time of arrival of a wavefront at the electrical centre of the two component antennas, then

$$\psi = 2\pi\nu_0\tau = 2\pi(ul+vm), \tag{5.8}$$

where ν_0 is the radio frequency and u and v wavelengths are the differences in the x and y directions between the antenna centres. For noise voltages with a finite frequency band $\Delta\nu$ Hz the cosine factor should be replaced by its average over all frequencies in the band. The difference between the exact formula and (5.7) (with ν_0 = band centre frequency) is negligible when

$$2\pi\Delta\nu\tau \ll 1. \tag{5.9}$$

We shall discuss here only this 'narrow-band' case. The 'bandwidth effects' which appear when this inequality is not satisfied will be discussed later. Substituting (5.5) and (5.6) into (5.7) we find

$$\left.\begin{aligned} p_c &= \text{const.}\,(p_1 p_2)^{\frac{1}{2}} \cos\psi, \\ A_c &= \text{const.}\,(A_1 A_2)^{\frac{1}{2}} \cos\psi. \end{aligned}\right\} \tag{5.10}$$

The value of the constant in these equations is derived in Appendix 3 by discussing in detail the impedance matching conditions and the definition of available power. The result is const. = 2 and the available correlated power per unit bandwidth becomes

$$p_c = 2(p_1 \cdot p_2)^{\frac{1}{2}} \cos \psi \qquad (5.11)$$

and the effective area (Figure 5.1 *a*) is

$$A_c = 2(A_1 \cdot A_2)^{\frac{1}{2}} \cos \psi. \qquad (5.12)$$

A_1, A_2 and the phase difference ψ will, in general, be functions of the direction from which the radio waves are coming. The effective area of the correlation telescope will, therefore, be written $A_c(l, m)$ when we want to stress its dependence on direction. Note that, in contrast to a total power telescope, both the power p_c and the effective area A_c can be negative as well as positive, depending on the value of the phase difference ψ.

The effective area can also be expressed in terms of the component antenna field patterns F_1 and F_2. The phase of a field pattern was defined as that of the output voltage relative to that of a reference voltage (section 2.3). The phase *difference* ψ must therefore be equal to the difference $(\phi_1 - \phi_2)$ between the field pattern phases in the direction from which the waves are coming; the reference phase drops out of the equation. Let

$$A_1 = A_{1\max} |F_1|^2 \quad \text{and} \quad A_2 = A_{2\max} |F_2|^2$$

and let the antenna beams be steered so that the directions of their maxima coincide. Equation (5.12) can then be written

$$A_c = 2(A_{1\max} A_{2\max})^{\frac{1}{2}} |F_1| \cdot |F_2| \cos(\phi_1 - \phi_2)$$
$$= 2(A_{1\max} A_{2\max})^{\frac{1}{2}} \cdot \mathrm{Re}\{F_1 F_2^*\}. \qquad (5.13)$$

The star indicates that the complex conjugate of F_2 should be used. The equation will, of course, be valid only when the phases of the two field patterns have been defined relative to the same reference voltage.

5.2.3. *Cosine and sine effective area*

The effective area $A_c(l, m)$ m² is called the *cosine effective area* of the correlation telescope, and the receiver discussed so far has been a *cosine receiver*, which measures the cosine correlated power p_c. The telescope is insensitive to radio waves from directions for which the phase difference ψ has the values given by

$$\psi = \pi/2 \pm i\pi \quad (i = 0, 1, 2 \ldots). \tag{5.14}$$

When ψ has these values, however, the telescope will respond if connected to a *sine receiver*. A cosine receiver can be changed into a sine receiver by adding an extra quarter period delay in front of one of the receiver inputs, as is shown in Figure 5.1*b*. The signals from

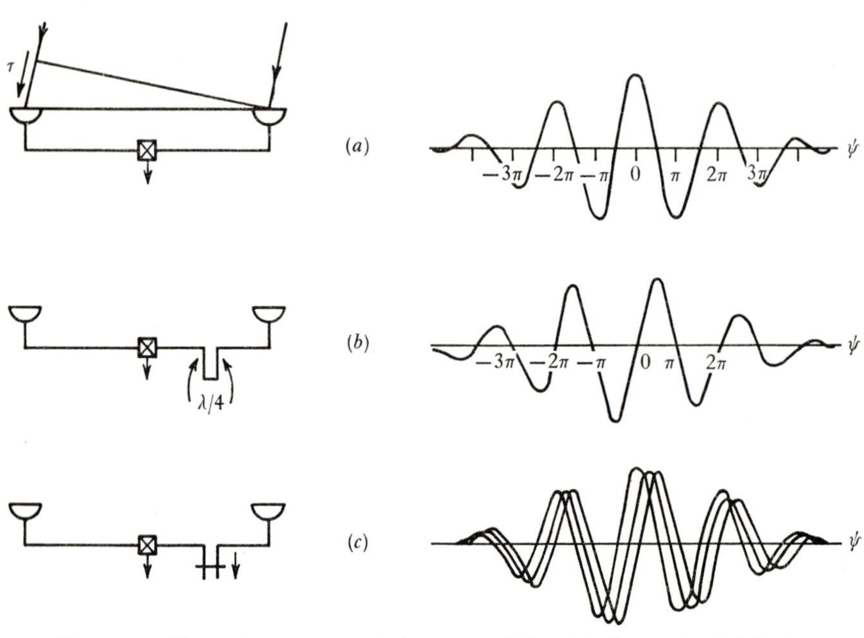

Figure 5.1. Two antennas connected as a correlation interferometer. (*a*) Cosine connection and the output from the correlator. (*b*) Sine connection and correlator output. (*c*) Variable-phase connection changing correlator output between sine and cosine modes. If the connection is made randomly, both sine and cosine components will usually appear in the output.

the second antenna will then be delayed by an extra quarter period, or $\pi/2$ radians of phase, and the phase difference between the voltages at the receiver inputs will have changed by the same amount. The telescope now has a different effective area $A_s(l, m)$, the *sine effective area*:

$$\begin{aligned} A_s &= 2(A_1 . A_2)^{\frac{1}{2}} \cos(\psi - \pi/2) \quad (\mathrm{m}^2) \\ &= 2(A_1 . A_2)^{\frac{1}{2}} \sin \psi \\ &= 2(A_{1\max} A_{2\max})^{\frac{1}{2}} . \operatorname{Im}\{F_1 F_2^*\} \end{aligned} \tag{5.15}$$

and the sine receiver measures the sine correlated power per unit bandwidth p_s

$$p_s = A_s S_m \quad (\mathrm{W\,Hz^{-1}}). \tag{5.16}$$

Correlation telescopes are often designed to measure both the cosine and the sine outputs by time sharing between periods when the extra $\lambda/4$ delay is connected and disconnected, or by using two receivers, one cosine and one sine, connected in parallel to the same two component antennas.

The effective area of the correlation telescope, when there is an arbitrary delay ϕ radians of r.f. phase in the one lead, will be (Figure 5.1*c*)

$$\begin{aligned} A_\phi &= 2(A_1 . A_2)^{\frac{1}{2}} \cos(\psi - \phi) \\ &= A_c \cos\phi + A_s \sin\phi \quad (\mathrm{m^2}) \end{aligned} \tag{5.17}$$

and it follows that

$$p_\phi = p_c \cos\phi + p_s \sin\phi \quad (\mathrm{W\,Hz^{-1}}). \tag{5.18}$$

The output from a telescope with any known phasing ϕ can therefore be calculated if the cosine and sine outputs have been measured and it will not contain any new information about the incoming radio signals.

5.2.4. *Complex form of the effective area: Envelope pattern*

The expressions for the cosine and sine effective areas (5.12), (5.13) and (5.15) show that these can be regarded as the real and the imaginary parts respectively of a complex effective area $\mathbf{A}(l, m)$:

$$\begin{aligned} \mathbf{A} &= A_c + jA_s \\ &= 2(A_1 A_2)^{\frac{1}{2}} \exp j\psi \\ &= |\mathbf{A}|_{\max} F_1 F_2^*, \end{aligned} \tag{5.19}$$

where

$$|\mathbf{A}|_{\max} = 2(A_{1\max} A_{2\max})^{\frac{1}{2}}. \tag{5.20}$$

The concept of a complex effective area will be used extensively in later sections of this book.

It is clear from these equations that the modulus $|\mathbf{A}(l, m)|$ of this complex effective area forms the envelope to the patterns with all possible phasings ψ (5.12)

$$|\mathbf{A}(l, m)| = 2[A_1(l, m) . A_2(l, m)]^{\frac{1}{2}}. \tag{5.21}$$

It follows from the inequality

$$2(A_1 . A_2)^{\frac{1}{2}} \leqslant A_1 + A_2 \tag{5.22}$$

that this *envelope pattern* $|\mathbf{A}(l, m)|$ is equal to or smaller than the sum of the two component antenna effective areas. It is equal to this sum in directions where $A_1 = A_2$ (which is the whole sky if the two antennas are identical). The envelope pattern equals the larger of the two effective areas if these are in the ratio 4:1, i.e.

$$|\mathbf{A}| = 2(A_1 . A_2)^{\frac{1}{2}} = A_1 \quad \text{(in the directions where } A_1 = 4A_2\text{)}. \tag{5.23}$$

The value of the envelope pattern for larger ratios A_1/A_2 is smaller than A_1 alone. The envelope pattern is clearly seen in Figure 5.1.

The total power telescope can formally be treated as a special case of a correlation telescope where the component antennas are identical and superposed ($F_1 \equiv F_2$), which, of course, is not possible in practice. However, (5.20) does not apply to total-power telescopes as it contains a factor 2 arising from the presence of *two separate non-interacting* antennas in an ideal correlation telescope.

5.2.5. *Correlation temperature*

In section 2.2 the antenna temperature T_a of a total-power telescope was defined as the temperature at which a resistor produces the same available power per unit bandwidth p as does the antenna:

$$p = kT_a \quad (\text{W Hz}^{-1}), \tag{2.11}$$

$k = 1{\cdot}38 \times 10^{-23}$ (W Hz^{-1} °K) is Boltzmann's constant. The antenna temperature of a lossless total-power telescope was derived in section 2.2:

$$T_a = \lambda^{-2} \int_{4\pi} T_B(l, m) . A(l, m) . d\Omega. \tag{2.14}$$

$T_B(l, m)$ is the sky brightness temperature distribution and $A(l, m)$ the effective area of the total-power telescope.

For a correlation telescope we define the *cosine* and *sine correlation temperatures* by equations analogous to (2.11):

$$\left.\begin{aligned} p_c &= kT_c \quad (\text{W Hz}^{-1}), \\ p_s &= kT_s. \end{aligned}\right\} \tag{5.24}$$

p_c and p_s are related to A_c and A_s respectively in exactly the same way as, for a total-power telescope, p is related to A (5.6), (5.16) and (2.1). The correlation temperatures of a lossless correlation telescope are, then, in direct analogy with (2.14):

$$\left.\begin{aligned} T_c &= \lambda^{-2}\int_{4\pi} T_B(l,m)\,A_c(l,m)\,d\Omega, \\ T_s &= \lambda^{-2}\int_{4\pi} T_B(l,m)\,A_s(l,m)\,d\Omega. \end{aligned}\right\} \qquad (5.25)$$

It will be convenient for the further discussion to condense these two equations into one single equation by regarding T_c and T_s as the real and imaginary parts respectively of a *complex correlation temperature* $\mathbf{T}_a$:

$$\mathbf{T}_a = T_c + jT_s. \qquad (5.26)$$

Then

$$\mathbf{T}_a = \lambda^{-2}\int_{4\pi} T_B(l,m)\,\mathbf{A}(l,m)\,d\Omega, \qquad (5.27)$$

where $\mathbf{A} = A_c + jA_s$ is the complex effective area (5.19). The expression for $\mathbf{T}_a$ is identical with that for a total-power telescope (2.14) except that, in (2.14), all quantities are real and positive.

5.2.6. *The all-sky integral of the correlation effective area*

In section 2.2 Equation 2.16 showed that the all-sky integral of the effective area of a total-power telescope without internal losses is equal to λ^2. A simple thermodynamical argument shows that the same integral for a correlation telescope is zero. The correlation receiver output can be expressed as the difference between two total power outputs, namely, when the two component antennas are connected together directly and when the phase of one is reversed before the connection.

This was the basis of the phase-switched correlation receiver described in section 5.1:

$$\overline{(V_1+V_2)^2} - \overline{(V_1-V_2)^2} = 4\overline{V_1 V_2}. \qquad (5.4)$$

Let T_a and T'_a be the total-power antenna temperatures of the combinations without and with, respectively, the inserted π radian phase change in the line to the second antenna. T_c is the correlation temperature when the two antennas are used with a correlation receiver. Then, from (5.25),

$$T_a - T'_a = \text{const. } T_c$$

$$= \text{const. } \lambda^{-2} \int_{4\pi} T_B(l,m)\, A_c(l,m)\, d\Omega, \tag{5.28}$$

where the value of the constant depends on the impedance matching conditions (see Appendix 3).

Imagine the whole antenna system placed in a black-body enclosure in thermal equilibrium at a temperature T°K. Then $T_B(l,m) = T$ in all directions and since, in thermal equilibrium, any total-power telescope must have the antenna temperature T°K we have

$$T_a = T'_a = T_B(l,m) = T \tag{5.29}$$

and (5.28) reduces to

$$\int_{4\pi} A_c(l,m)\, d\Omega = 0. \tag{5.30}$$

The same argument applies if we add a $\pi/2$ phase shift so that the receiver becomes a sine receiver. We can therefore derive a more general equation for $\mathbf{A} = A_c + jA_s$

$$\int_{4\pi} \mathbf{A}(l,m)\, d\Omega = 0. \tag{5.31}$$

The difference between the all-sky effective area integrals for total-power and correlation telescopes shows that the reception pattern of a correlation telescope can never have exactly the same shape as that of a total-power telescope. This does not hold for 'mixed' systems; it is possible to add a certain amount of total-power output to the normal output of a correlation telescope and in this way make the beam similar to that of a total-power telescope (see Chapter 6).

5.2.7. *Making use of the product pattern: The cross telescope*

The product of two field patterns (5.19) will, in general, be large only in those directions which are common to the main parts of both patterns. This common region can be much smaller than the individual patterns if these are very different in shape. The well-known *cross telescope* is an example of a telescope which makes use of this principle to achieve a high resolving power.[502]

The cross telescope consists of two orthogonal long narrow-aperture antennas and the telescope has a narrow 'pencil' response where the corresponding two orthogonal fan-shaped patterns cross in the sky (Figure 5.2).

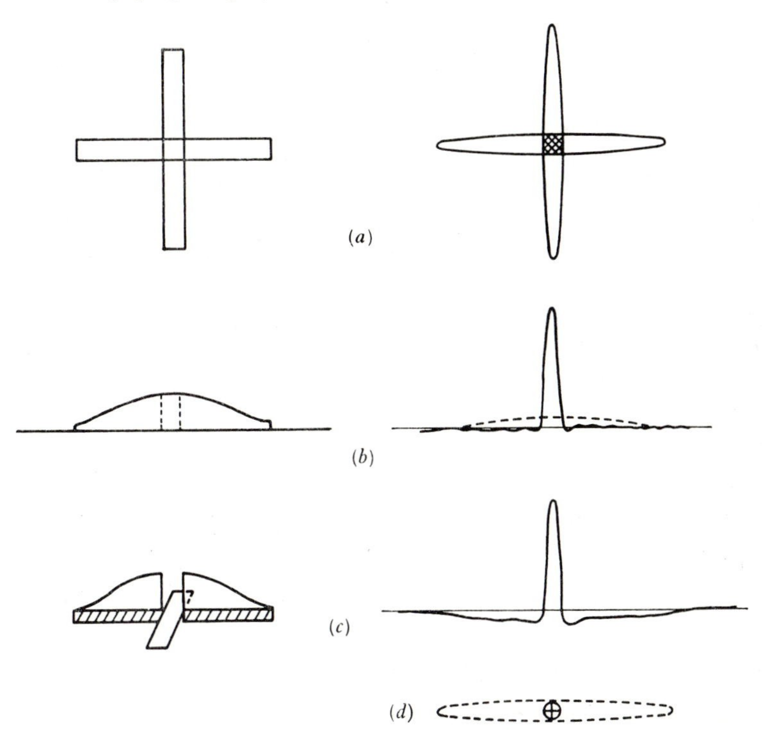

Figure 5.2. (*a*) A cross telescope and the field patterns (right) of the two parts of the cross. (*b*) The grading of one array of the cross and its field pattern. The dashed line indicates the field pattern of the central part alone. (*c*) The grading of one array with the central part removed. The field pattern is that of (*b*) minus the field pattern of the central portion of the cross. (*d*) When the two arrays of the cross are multiplied a pencil beam is formed. It is accompanied by an extensive weak negative sidelobe.

Let the antennas be L wavelengths long and W wavelengths wide. The field patterns of the two rectangular-aperture cross arms are (2.44):

$$\left.\begin{aligned} F_1(l,m) &= F_e(l,m)\,.\,\frac{\sin(\pi Ll)}{\pi Ll}\,.\,\frac{\sin(\pi Wm)}{\pi Wm}, \\ F_2(l,m) &= F_e(l,m)\,.\,\frac{\sin(\pi Wl)}{\pi Wl}\,.\,\frac{\sin(\pi Lm)}{\pi Lm}. \end{aligned}\right\} \tag{5.32}$$

Usually $L \gg W$ and the factors containing W as well as the element field pattern F_e can be approximated by 1·0 over the main part of the *product* pattern $F_1 F_2^*$. The cosine and sine effective areas are then

$$A_c(l, m) = |\mathbf{A}|_{\max} \cdot R_e(F_1 F_2^*)$$
$$= |\mathbf{A}|_{\max} \cdot \frac{\sin(\pi L l)}{\pi L l} \cdot \frac{\sin(\pi L m)}{\pi L m}. \qquad (5{\cdot}33)$$

$$A_s(l, m) = |\mathbf{A}|_{\max} \cdot \mathrm{Im}\,(F_1 F_2^*) = 0. \qquad (5{\cdot}34)$$

The *power response* (cosine effective area) of the cross telescope is similar to the *field pattern* of a $L \times L$ wavelength² filled uniform aperture (2.44). It has rather prominent sidelobes, but these can be reduced to any desired level by increasing the length of the two arrays and giving them tapered instead of uniform gradings (Figure 2.7). The beamwidth of a tapered cross will be roughly comparable with that of a filled square aperture with sides equal to half the length L of the cross arms. The maximum effective area $|\mathbf{A}|_{\max}$, on the other hand, is smaller and approximately proportional to the geometrical areas of the two telescopes; the exact value will depend on the grading of the antennas. The formation of negative side lobes (Figure 5.2) to satisfy the requirement for a zero all-sky integral of the effective area A_c, will be discussed in Chapter 6.

5·3. The telescope transfer function

5·3·1. *Introducing the transfer function*

It is often convenient to describe the correlation telescope beam as the product of two field patterns, as we did in section 5.2. This is, however, less helpful for more complex antenna arrangements and, in addition, it does not give a very clear indication of the general relations between the antenna structure and the resulting telescope beam. The *telescope transfer function* (sometimes called the spatial spectral sensitivity function) is a very useful tool for dealing with and comparing radio telescopes. We shall usually assume that all antennas are very large in both dimensions, compared with a wavelength, so that the current element pattern F_e will be ignored. Starting from (5.19) in its one-dimensional form

$$\mathbf{A}(l) = |\mathbf{A}|_{\max} \cdot F_1(l) \, . \, F_2^*(l). \qquad (5{\cdot}35)$$

Writing the two field patterns as proportional to the Fourier transforms of the corresponding gradings $g_1(x)$ and $g_2(x)$, and collecting all the proportionality constants together we have

$$\mathbf{A}(l) = \text{const.}\int g_1(x)\exp\{+j2\pi xl\}\,dx\,.\int g_2^*(x')\exp\{-j2\pi x'l\}\,dx'$$

$$= \text{const.}\iint g_1(x)\,.\,g_2^*(x')\exp\{j2\pi(x-x')\,l\}\,dx\,dx'. \qquad (5.36)$$

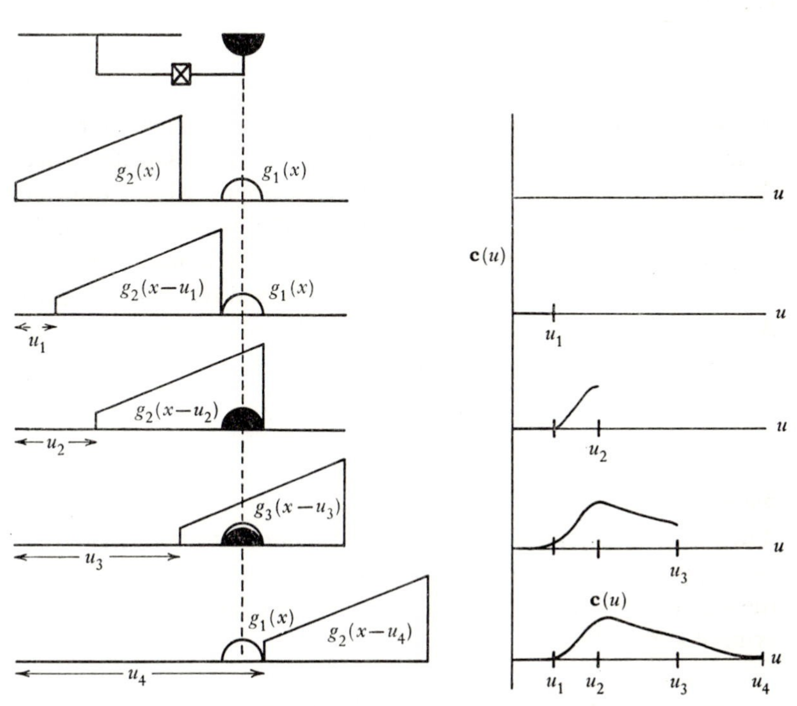

Figure 5.3. The construction of the *transfer function*, or spatial spectral sensitivity function $\mathbf{c}(u)$ of a correlation telescope by smoothing the grading of one antenna by the grading (more generally, the complex conjugate of the grading) of the other.

The double integral is to be taken over all combinations (x, x') of the aperture elements, one at x in antenna no. 1 and one at x' in antenna no. 2. We now introduce the spacing u between such a pair of aperture elements:

$$u = x - x' \qquad (5.37)$$

and eliminate x' from (5.36). Then

$$\mathbf{A}(l) = \int\left[\text{const.}\int g_1(x)\,.\,g_2^*(x-u)\,dx\right]\exp\{j2\pi ul\}\,du. \quad (5.38)$$

The expression within the square brackets is the *telescope transfer function* $\mathbf{c}(u)$.

$$\mathbf{c}(u) = \text{const.}\int g_1(x)\,.\,g_2^*(x-u)\,dx. \quad (5.39)$$

Hence

$$\mathbf{A}(l) = \int \mathbf{c}(u)\exp\{j2\pi ul\}\,du \quad (5.40)$$

or, symbolically,

$$\mathbf{A}(l) \mathscr{F} \mathbf{c}(u). \quad (5.41)$$

Thus the effective area of a telescope is the Fourier transform of its transfer function. The effective area is completely specified once the transfer function is known, and telescopes whose transfer functions are similar have similar beams. The transfer function is usually normalized so that its modulus has a maximum value of unity. The Fourier transform of this normalized function will give $\mathbf{A}(l)$, apart from a scaling factor which can be calculated with the aid of the equation already derived,

$$|\mathbf{A}|_{\max} = 2(A_{1\max}A_{2\max})^{\frac{1}{2}}. \quad (5.20)$$

5.3.2. *Constructing transfer functions*

The integral in (5.39) is known as the *smoothing integral* of g_1 and g_2^* (see Appendix 2) and differs from the convolution integral only in the sign of $(x-u)$. The smoothing integral will be indicated by the symbol $\star$.[503–4] Hence

$$\mathbf{c}(u) = \text{const.}\int g_1(x)\,.\,g_2^*(x-u)\,dx \equiv \text{const.}\,g_1(x)\star g_2^*(x). \quad (5.42)$$

The smoothing integral has a relatively simple geometrical meaning which makes it easy to compare the performance of different radiotelescopes with respect to their angular resolution, beam shape, sidelobes, etc., by comparing their transfer functions.

Equation (5.42) shows that $\mathbf{c}(u)$ equals the integral of the product $g_1g_2^*$ when g_2^* has been shifted an amount u from its original place. Figure 5.3 shows the transfer function for a correlation telescope consisting of two dissimilar component antennas (compare the construction of the convolution integral in Figure 2.8).

The region covered by the transfer function along the u-axis represents the total range of available spacings u between aperture elements in the two antennas; g_1 and the shifted g_2^* will overlap and $\mathbf{c}(u)$ be different from zero only if there exist, in the original configuration, two aperture elements, one in each antenna, which are separated by u wavelengths. The modulus and argument of $\mathbf{c}(u)$ express the amplitude product and the phase difference respectively of the resultant of all pairs of aperture elements which are spaced by u wavelengths.

The transfer function in two dimensions $\mathbf{c}(u, v)$ relates to pairs of aperture elements which are separated by u wavelengths in the x direction and v wavelengths in the y direction. The relations (5.41) and (5.42) in two dimensions become

$$\mathbf{A}(l, m) \mathscr{F} \mathbf{c}(u, v)$$

$$\mathbf{c}(u, v) = \text{const.}\, g_1(x, y) \star g_2^*(x, y). \tag{5.43}$$

Figure 5.4 shows an example of a two-dimensional transfer function ('spatial spectral sensitivity diagram').

5.3.3. *Total-power telescopes*

A total-power effective area can be written in the same form as (5.35) but with identical field patterns. Consequently, the transfer function appears as a smoothing integral of identical gradings (auto-correlation function)

$$\mathbf{c}(u, v) = \text{const.}\, g(x, y) \star g^*(x, y). \tag{5.44}$$

Figure 5.5 shows examples of total-power transfer functions. The available spacings u are spread symmetrically about the origin at which $\mathbf{c}$ has its maximum value. This contrasts with its form for a correlation telescope where the zero spacing is absent and therefore $\mathbf{c}(0, 0) = 0$. This reflects the difference in the all-sky integrals of the effective areas (2.16) and (5.31). Let the beam be narrow in both dimensions and centred at the direction (l_0, m_0); then

$$d\Omega = dl\,dm/(1 - l^2 - m^2)^{\frac{1}{2}} \simeq dl\,dm/(1 - l_0^2 - m_0^2)^{\frac{1}{2}}$$

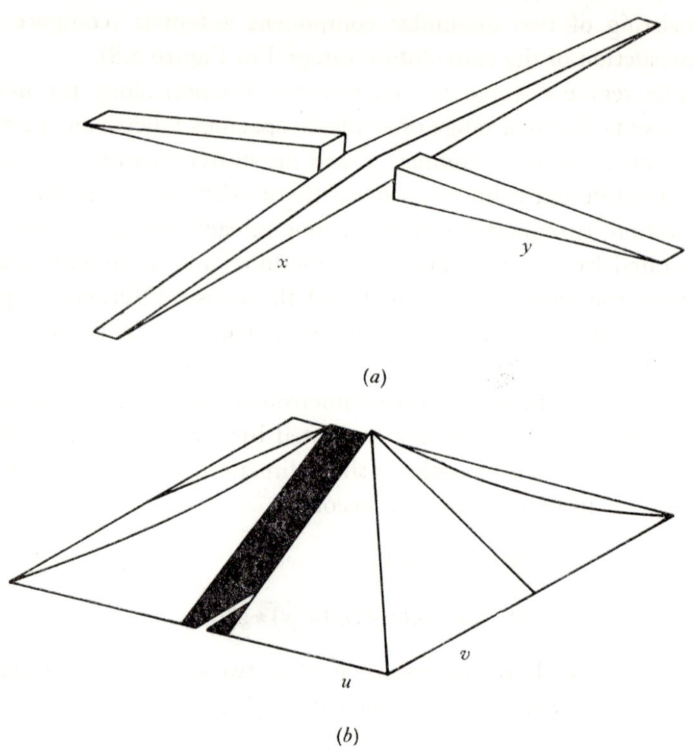

Figure 5.4. (*a*) The gradings $g_1(x, y)$ and $g_2(x, y)$ of a tapered cross telescope. (*b*) The transfer function $\mathbf{c}(u, v)$ of the cross telescope.

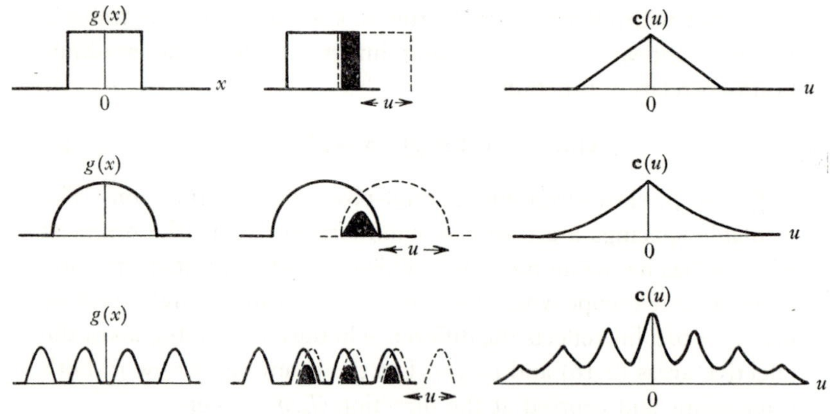

Figure 5.5. The total-power transfer functions $\mathbf{c}(u)$ corresponding to several gradings $g(x)$. $\mathbf{c}(u)$ is obtained as a smoothing integral of a grading by itself (or more generally by its complex conjugate). It is sometimes called the autocorrelation function.

over the main part of the pattern. The functions $\mathbf{c}(u, v)$ and $\mathbf{A}(l, m)$ constitute a Fourier pair, thus:

$$\mathbf{c}(0,0) = \iint \mathbf{A}(l,m)\, dl\, dm \approx (1 - l_0^2 - m_0^2)^{\frac{1}{2}} \iint \mathbf{A}(l,m)\, d\Omega. \quad (5.45)$$

$\mathbf{c}(0, 0)$ will thus be finite or zero, depending on whether we are dealing with a total-power or a correlation telescope.

5.4. Measuring the sky brightness distribution with a radio-telescope: Antenna smoothing

The correlation temperature (or for a total power telescope the antenna temperature) caused by a particular sky brightness temperature distribution $T_B(l, m)$ was derived in section 5.2.5:

$$\mathbf{T}_a = \lambda^{-2} \int_{4\pi} T_B(l,m)\, \mathbf{A}(l,m)\, d\Omega. \quad (5.27)$$

Let $\mathbf{A}'(l, m)$ be the telescope effective area when the beam has steered to point in the direction (l_0, m_0). In practice we are concerned with the measured correlation (or antenna) temperature $\mathbf{T}_a$ as a function of the direction (l_0, m_0) to which the beam has been steered. Hence we shall write

$$\mathbf{T}_a(l_0, m_0) = \lambda^{-2} \int_{4\pi} T_B(l,m)\, \mathbf{A}'(l,m)\, d\Omega. \quad (5.46)$$

5.4.1. *Electrical steering*

We showed in section 2.4.4 (eqn 2.57) that a change in an antenna grading from $g(x)$ to $g(x) . \exp\{j2\pi sx\}$—i.e. a phase change linear in x—will shift the Fourier transform of the grading and, therefore, also the field and power patterns by an amount s in the coordinate l. Similarly, a phase shift $2\pi(l_0 x + m_0 y)$ radians applied to an antenna with the two-dimensional grading $g(x, y)$ will shift its power pattern P from the l, m origin direction to the direction l_0, m_0 and the shifted power pattern can be written

$$P'(l,m) = P(l - l_0, m - m_0). \quad (5.47)$$

The shifted and unshifted power patterns have the same shape when referred to the particular l, m coordinate system, but they will, of course, differ when expressed in angular measure. The all-sky integral of the effective area of any total power antenna must be equal to λ^2 or, if internal losses are present, $\eta_R \lambda^2$. Hence

$$\int_{4\pi} A_{\max} P(l,m)\,d\Omega = \int_{4\pi} A'_{\max} P'(l,m)\,d\Omega = \lambda^2, \tag{5.48}$$

where

$$d\Omega = dl\,dm/(1-l^2-m^2)^{\frac{1}{2}}. \tag{2.5}$$

If the antenna pattern occupies a small region of the sky, then we can approximate the square-root factor by its value at the centre of the beam, i.e. in the first integral by 1·0 and in the second integral by $(1-l_0^2-m_0^2)^{\frac{1}{2}}$. Hence

$$A_{\max} \iint P(l,m)\,dl\,dm = A'_{\max}(1-l_0^2-m_0^2)^{-\frac{1}{2}} \iint P'(l,m)\,dl\,dm. \tag{5.49}$$

It follows from (5.47) that the two integrals are identical as long as the main contributing parts of the patterns are well within the area of integration $l^2+m^2 \leqslant 1$. Then

$$\begin{aligned} A'_{\max} &= A_{\max}(1-l_0^2-m_0^2)^{\frac{1}{2}} \\ &= A_{\max}\cos\gamma_0, \end{aligned} \tag{5.50}$$

where γ_0 is the angle between the directions $(0, 0)$ and (l_0, m_0). The equation simply states that the maximum effective area of the shifted beam is proportional to the geometrical area seen in projection from the direction of the beam. Note that this does not hold, for example, for a line antenna, since its pattern is not confined to a small region of sky and the approximations that were used in deriving this equation will not be justified.

The same equation (5.50) will clearly be valid for a correlation telescope if it is valid, separately, for the two component antennas and these are both steered electrically in the direction (l_0, m_0). Hence we can write, more generally,

$$\mathbf{A}'(l,m) = \mathbf{A}(l-l_0, m-m_0)\,.\,(1-l_0^2-m_0^2)^{\frac{1}{2}}. \tag{5.51}$$

We substitute this expression for A' into (5.46). Replacing, again, the square root factor by its value at the centre of the beam, $(1-l_0^2-m_0^2)^{\frac{1}{2}}$, we get

$$\begin{aligned} \mathbf{T}_a(l_0,m_0) &= \lambda^{-2} \iint_{-\infty}^{+\infty} T_B(l,m)\,.\,\mathbf{A}(l-l_0, m-m_0)\,dl\,dm \\ &= \lambda^{-2}\,.\,[T_B(l_0,m_0) \star \mathbf{A}(l_0,m_0)] \end{aligned} \tag{5.52}$$

which shows formally that the measured brightness distribution $\mathbf{T}_a$ equals λ^{-2} times the true brightness distribution T_B smoothed by the effective area of the unshifted beam. The integration limits have been formally extended to $\pm\infty$ which is permitted if we make the logical definition that $T_B \equiv 0$ outside the circular area $l^2+m^2 \leqslant 1$ that represents the celestial sphere.

5.4.2. *Mechanical steering*

The beam of a mechanically steered telescope is shifted relative to the sky by tilting the whole antenna structure mechanically. A stationary telescope used for 'drift scan' measurements can be regarded as a special case of a mechanically steered telescope: the whole antenna structure rotates with the Earth while the measurements are being made. Clearly, for a mechanically steered telescope, the beam retains its shape when expressed in *angular* measure (apart from changes due, for example, to gravitational deformation of the antenna structure when it is tilted). Hence, in contrast to the electrically steered telescope, the beam will not keep its shape when described in the 'projected' coordinates l, m. However, for l and $m \ll 1$ these coordinates approximate angles in radians and, with the extra restriction that the shift (l_0, m_0) must be much smaller than one radian, we can use (5.52) for a mechanically steered telescope.

5.4.3. *Antenna smoothing in terms of the telescope transfer function*

Equation (5.52) shows in a direct way the effect of the telescope reception pattern on the measured sky brightness distribution. There is another way of describing the same phenomenon which will be useful in later discussions, particularly in connection with synthesis radiotelescopes.

A direct Fourier transform followed by an inverse Fourier transform of the result will restore the original function. Hence, a function can be written as the inverse transform of its own Fourier transform. Applying this to (5.52) we can write, after dropping the indices of l_0 and m_0.

$$\mathbf{T}_a(l,m) \mathscr{F} \lambda^{-2}[\text{Fourier transform of } T_B(l,m) \star \mathbf{A}(l,m)]. \quad (5.53)$$

Let $\mathbf{t}(u,v)$ be the Fourier transform of $T_B(l,m)$. Then (see Appendix 2):
$$\mathbf{T}_a(l,m) \mathscr{F} \lambda^{-2}[\mathbf{t}(u,v) \,.\, \mathbf{c}(u,v)] \quad (5.54)$$

$\mathbf{c}(u, v)$ is, as before, the telescope transfer function, i.e. the inverse Fourier transform of $\mathbf{A}(l, m)$ (5.43). The smoothing of the true distribution $T_B(l, m)$ by the beam $\mathbf{A}(l, m)$ is therefore equivalent to multiplying its Fourier transform $\mathbf{t}(u, v)$ by the telescope transfer

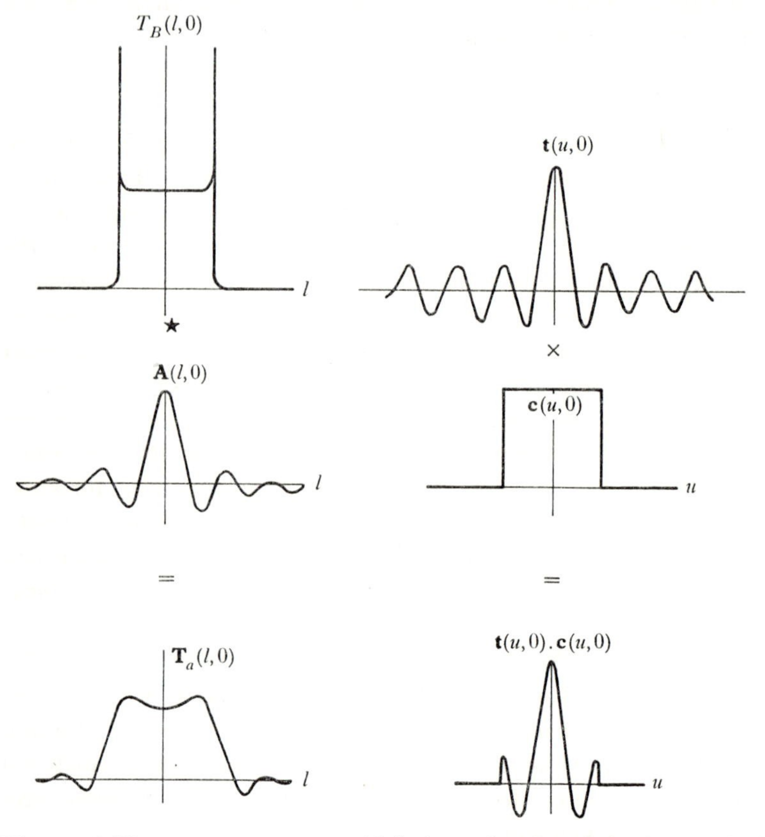

Figure 5.6. The antenna temperature $\mathbf{T}_a(l, 0)$ as a function of the direction l of the beam centre is the true brightness temperature of the source $T_B(l, 0)$ smoothed by the effective area $\mathbf{A}(l, 0)$ of the antenna. When the process is expressed in terms of the transform $\mathbf{t}$ of T_B and the inverse transform $\mathbf{c}$ of $\mathbf{A}$, then the smoothing is replaced by a simple multiplication.

function $\mathbf{c}(u, v)$, as Figure 5.6 illustrates. The transfer function will be zero everywhere except in the limited u, v region that represents the available antenna element spacings in the telescope (section 5.32). Since only the product $[\mathbf{t} . \mathbf{c}]$ can be derived from the measurements, we shall have an incomplete knowledge of $\mathbf{t}$ and, consequently, of

T_B. The restricted knowledge of **t** and the smoothed or 'blurred' picture of T_B are mathematically the same thing: a certain amount of information has been lost and cannot be retrieved by any mathematical operation on the product $[\mathbf{t}.\mathbf{c}]$ or on the smoothed map $\lambda^{-2}[T_B \star \mathbf{A}]$. It is, however, possible to present the available information in different ways corresponding to measurements taken with different 'possible' beams. This process is known as 'restoring' and will be described in section 6.23.

5.5. Broad-band systems

The discussion of antennas has been confined, up to this point, to the performance over a range of wavelengths so narrow that the antenna characteristics can be considered to be the same for any wavelength in the band accepted. For many radiotelescopes this assumption involves little error, but for the larger systems which we shall be describing in the following chapters it is necessary to take into account the effects on the angular response of the telescope of the finite width of the band of wavelengths accepted by the receiver.

A radiotelescope in which the same time is taken for an impulse to travel from the source, through any part of the antenna, to the receiver is suitable for reception over a wide range of frequencies. A parabolic reflecting telescope has this characteristic for radiation arriving at the telescope from the direction of the axis of the paraboloid. On the other hand, a widely spaced pair of antennas used as an interferometer and receiving energy from a source that is not in a direction normal to the line of the interferometer has, unless special precautions are taken, a reception pattern that is strongly dependent on the exact frequency of observation. If a broad-band receiver is used then the different interference patterns of different frequencies inside the reception band will add up to a pattern which may differ greatly from that in a narrow-band receiver.

5.5.1. *Broad-band telescope systems*

If $\mathbf{A}_\nu(l, m)$ is the effective area at frequency ν, then the *broad-band effective area* is given by the weighted average of the effective area over all frequencies in the band.

$$\mathbf{A}_B(l, m) = \int_0^\infty \mathbf{A}_\nu(l, m)\, b(\nu)\, d\nu, \tag{5.55}$$

$b(\nu)$ is the bandpass characteristic (spectral sensitivity) of the receiver and normalized so that

$$\int_0^\infty b(\nu)\,d\nu = 1.$$

We can express $A_\nu(l, m)$ in terms of the effective area at the centre frequency ν_0. The field patterns at frequencies ν and ν_0 are proportional to the Fourier transforms of the antenna gradings at the two frequencies. Antennas are usually designed so that the shape of the grading does not change significantly over the frequency band to be used. The scale, however, does change since the grading is expressed in a coordinate system where the unit of length is one wavelength. Let x, y be the coordinate system at the frequency ν. Then

$$g_\nu(x, y) = g_{\nu_0}(x', y'), \tag{5.56}$$

where

$$(x', y') = (\nu_0/\nu)(x, y). \tag{5.57}$$

Hence

$$g_\nu(x, y) = g_{\nu_0}\left(\frac{\nu_0}{\nu}x, \frac{\nu_0}{\nu}y\right) \tag{5.58}$$

and from the similarity theorem (Appendix 2)

$$F_\nu(l, m) = F_{\nu_0}\left(\frac{\nu}{\nu_0}l, \frac{\nu}{\nu_0}m\right) \tag{5.59}$$

both patterns being normalized.

From this we see that

$$\mathbf{A}_\nu(l, m) = \frac{|\mathbf{A}_\nu|_{\max}}{|\mathbf{A}_{\nu_0}|_{\max}} \cdot \mathbf{A}_{\nu_0}\left(\frac{\nu}{\nu_0}l, \frac{\nu}{\nu_0}m\right) \tag{5.60}$$

For large antennas the maximum effective area depends only on the geometrical area and the form or shape of the grading (section 3.1, eqn 3.5). Hence $|\mathbf{A}_\nu|_{\max} \approx |\mathbf{A}_{\nu_0}|_{\max}$ and we get

$$\mathbf{A}_\nu(l, m) = \mathbf{A}_{\nu_0}\left(\frac{\nu}{\nu_0}l, \frac{\nu}{\nu_0}m\right). \tag{5.61}$$

This equation shows that the effective area (and the field pattern) has the same shape at all frequencies within the band, but the scale changes with frequency: it shrinks or expands about the direction $l = 0$, $m = 0$ as the frequency is increased or decreased respectively. The introduction of delays into different parts of the system will modify this equation. We shall deal with this in section 6.32.

If (5.61) is valid, separately, for both component antennas in a correlation telescope, it will clearly also be valid for the (complex) correlation effective area **A**. We can now write the broad-band effective area (5.55) as

$$\mathbf{A}_B(l, m) = \int_0^\infty \mathbf{A}_{\nu_0}\left(\frac{\nu}{\nu_0} l, \frac{\nu}{\nu_0} m\right) b(\nu)\, d\nu. \tag{5.62}$$

The effect of a change of frequency on the directional response of the antenna is most marked for systems in which the main response of the antenna is steered electrically from the direction $l = 0, m = 0$ so that the signal paths (time delays) from the source to the receiver are very different for signals that travel through different parts of the antenna. However, even for a mechanically steered antenna the beamwidth may be markedly different at the two ends of the reception band, because the size of the telescope aperture expressed in wavelengths is different for different wavelengths. The integration over the whole band will, however, make the broad-band effective area closely similar to the effective area at the centre frequency in many cases, e.g. when the telescope has a simple single beam directed at $l = 0$, $m = 0$. For an interferometer, however, where the pattern has fine structure in the fringes, the difference between broad-band and narrow-band patterns are often very great.

5.5.2. *A simple correlation interferometer*

Let the interferometer consist of two symmetrical antennas whose patterns both are centred at the zenith ($l = 0$, $m = 0$). Then its effective area at the frequency ν is

$$\mathbf{A}_\nu(l, m) = |\mathbf{A}_\nu(l, m)| \exp(j2\pi\nu\tau), \tag{5.63}$$

where τ is the time difference of the arrival of a wave front at the two antennas. We can, further, replace ν by $(\nu_0 + \nu')$, where ν' is now the frequency with respect to the centre of the band of reception. The spectral sensitivity in the receiving band will here be written $b'(\nu')\,[\equiv b(\nu_0 + \nu')]$. From (5.55)

$$\mathbf{A}_B(l, m) = \int_{-\nu_0}^\infty |\mathbf{A}_\nu(l, m)|\, b'(\nu') \exp\{j2\pi(\nu_0 + \nu')\tau\}\, d\nu'. \tag{5.64}$$

Approximating the smooth envelope pattern by its form at the centre frequency we get

$$\mathbf{A}_B(l,m) \simeq |\mathbf{A}_{\nu_0}(l,m)| \exp(j2\pi\nu_0\tau) \, . \int_{-\nu_0}^{\infty} b'(\nu') \exp(j2\pi\nu'\tau)\, d\nu'$$
$$= \mathbf{A}_{\nu_0}(l,m) \, . \, B(\tau), \tag{5.65}$$

where $B(\tau)$ is the Fourier transform of the bandpass $b'(\nu')$. The integral should have had the limits $\pm\infty$ but the formal extension of the lower limit to $-\infty$ is allowed if we define

$$b'(\nu') \equiv 0 \quad \text{for} \quad \nu' < -\nu_0.$$

The same result could have been obtained by applying (5.62). As an example we take a (normalized) rectangular pass band

$$b'(\nu') \begin{cases} = \dfrac{1}{\Delta\nu} & \text{for} \quad -\Delta\nu/2 < \nu' < \Delta\nu/2, \\ = 0 & \text{outside these limits.} \end{cases} \tag{5.66}$$

Then the envelope or 'fringe washing' function is

$$B(\tau) = \frac{\sin(\pi\Delta\nu\tau)}{\pi\Delta\nu\tau} \tag{5.67}$$

which is of the well-known form $\sin a/a$. Let the interferometer spacing be X *metres* in the x-direction; then $\tau = Xl/c$ and the first zeros of the 'fringe-washing' function $B(\tau) = B(Xl/c)$ will fall at $\Delta\nu\tau = 1$, i.e.

$$l = \pm \frac{c}{X\Delta\nu}. \tag{5.68}$$

This envelope to the angular response, produced by the broad-band response of the receiver, may reduce considerably the range of angles over which the interferometer can receive signals. In the extreme case $\Delta\nu = 2\nu_0$ when the band extends from $\nu = 0$ to $\nu = 2\nu_0$, all but the central 'white' fringe disappears. Examples are shown in connection with a later discussion in section 6.3.2 (Figure 6.21).

CHAPTER 6

UNFILLED-APERTURE ANTENNAS

In Chapter 1 we briefly discussed telescopes in which only part of the overall aperture is filled by the antenna structure. These telescopes make use of the fact that there is a great deal of redundancy in the information provided by a filled aperture and that the same information can be secured with a considerably reduced antenna system. Furthermore, it may happen that we already know something about the source (e.g. we may know that it is strong, isolated and small) and consequently we may require only a limited amount of additional information about it. This may allow us to use an incomplete antenna system.

In this chapter we shall discuss unfilled-aperture telescopes which contain, in themselves, all the necessary 'spatial components' (i.e. spacings u, v between aperture elements) for the measurement to be made. Aperture synthesis telescopes, in which the spatial components are obtained sequentially, we shall leave until Chapter 7.

6.1. Line apertures

6.1.1. *The grating antenna*

This antenna (array) is an example of an unfilled-aperture line antenna. In its simplest form it consists of a regular array of N identical antennas with a spacing d wavelengths between adjacent units and it is operated as a total power telescope (Figure 6.1 *a* and Plate 9, facing p. 96).[601] The missing parts of the filled line would contribute spatial components (spacings) which, as we shall see, are not necessary for observing isolated radio sources whose angular dimensions are less than $1/d$ radians. The remaining spacings, with the exception of the largest spacing $(N-1)d$, are however, still present with considerable redundancy, as Figure 6.1*c* shows.

Field pattern and array grading. The field pattern of a line antenna oriented along the x-axis is given by (2.32):

$$F(l,m) = \text{const.} F_e(l,m) \,.\, f(l), \qquad (2.32)$$

where $f(l)$ is the Fourier transform of the one-dimensional grading and the constant is to be chosen so that $|F(l,m)|_{max} = 1$. In the grating antenna we can let $F_e(l,m)$ represent the field pattern of an individual unit in the array; $f(l)$ will then be the Fourier transform of the 'array grading'—i.e. a set of spikes whose height g_i is proportional to the current carried by the unit number i. The Fourier transform is replaced by a Fourier series and

$$f(l) = \sum_{i=1}^{N} g_i \exp[j2\pi x_i l]. \tag{6.1}$$

The imaginary component of $f(l)$ will cancel if, for simplicity, we choose the origin at the array centre of symmetry and the exponential factor can then be replaced by $\cos(2\pi x_i l)$.

If the array is uniformly graded, i.e. if all $g_i = 1$, then the solution of (6.1) is

$$f(l) = N\frac{\sin(N\pi d \times l)}{N\sin(\pi d \times l)} \tag{6.2}$$

which, apart from the subscript x is identical with (4.4), which was derived in a different way. The field pattern of a grating antenna with a uniform array grading is thus

$$F(l,m) = F_e(l,m)\frac{\sin NU}{N\sin U}, \tag{6.3}$$

where $$U = \pi d \times l.$$

The field pattern has been normalized in the maximum direction. This maximum direction is not necessarily $l = 0$, $m = 0$, since the response of the unit antenna $F_e(l, m)$ depends on the direction in which it is pointed. Note that $F_e(l,m)$ is a variable response expressed in terms of the fixed coordinates (l,m) and it should not be confused with the constant field pattern $F_e(l',m')$ of the unit antenna, where l' and m' are directional cosines measured with respect to the direction of maximum response (l_0, m_0)—i.e. the pointing direction of the unit antennas.

The *effective area* of the grating antenna is given by the standard equation

$$A(l,m) = A_{max}|F(l,m)|^2. \tag{2.19}$$

We can express A_{max} directly in terms of the maximum effective area of the individual array units, $(A_e)_{max}$. In section 3.1 we derived a relation between $A_{\max}$ and the antenna grading $g(x,y)$:

$$A_{\max} = \frac{\lambda^2 \left| \iint g(x,y)\,dx\,dy \right|^2}{\iint |g(x,y)|^2\,dx\,dy} \tag{3.5}$$

which is valid for any antenna without internal losses whose pattern, like that of the reflector antennas for which it was derived, is centred at the direction $l = l_0$, $m = m_0$. Let $g_e(x,y)$ be the grading of a single array unit, g_i the array grading of unit number i and let there be no significant coupling between the currents in adjacent units. Then (3.5) can be written

$$A_{\max} = \frac{\lambda^2 \left| \iint g_e(x,y)\,dx\,dy \right|^2 \left[\sum_{i=1}^{N} g_i \right]^2}{\iint |g_e(x,y)|^2\,dx\,dy \,.\, \sum_{i=1}^{N} g_i^2}$$

$$= (A_e)_{\max} \frac{\left[\sum_{i=1}^{N} g_i \right]^2}{\sum_{i=1}^{N} g_i^2}. \tag{6.4}$$

Note that this equation will not be valid for closely spaced dipole arrays, since the individual dipoles are here strongly coupled to one another through their mutual impedances.

For a grating antenna with a uniform array grading (all $g_i = 1$) we find, as we expected, that

$$A_{\max} = N\,.\,(A_e)_{\max}. \tag{6.5}$$

Equations 6.3, 2.19, and 6.5 give its effective area

$$A(l,m) = N\,.\,A_e(l,m)\,.\left\{\frac{\sin NU}{N \sin U}\right\}^2, \tag{6.6}$$

where $U = \pi d \times l$. Figure 6.1 shows the pattern of such a uniform array. Maxima occur when $\sin U = 0$, i.e. when

$$\pm \quad l = k/d \quad (|k| = 0, 1, 2, 3, \ldots). \tag{6.7}$$

The angular distance between adjacent grating beams is $1/d$ in the

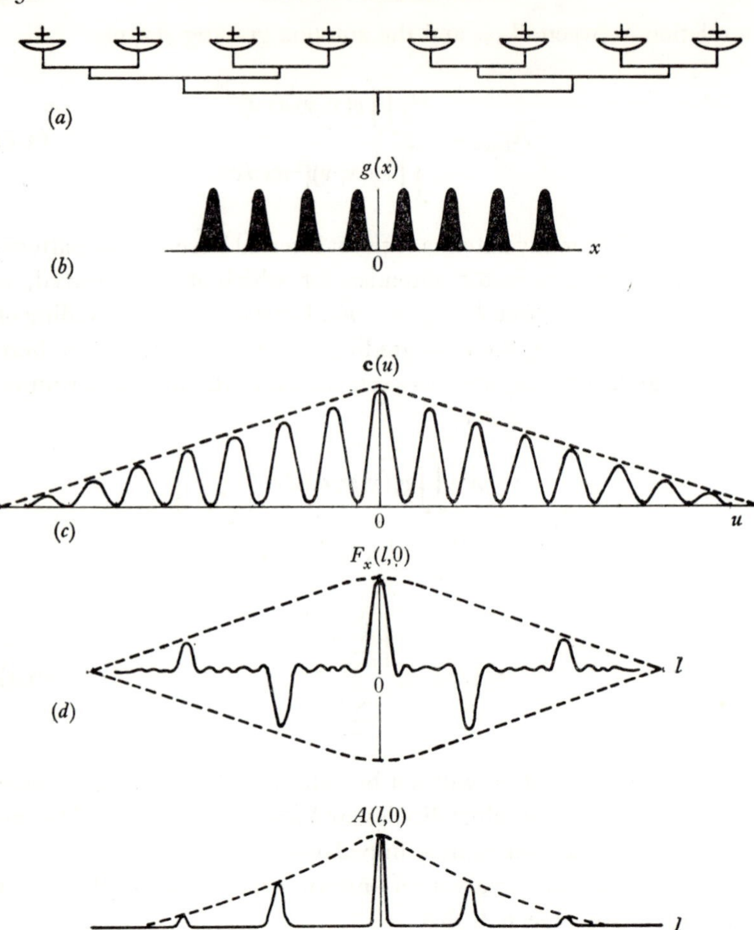

Figure 6.1. A grating telescope used as a total-power instrument. (*a*) General arrangement. (*b*) Grading $g(x)$. (*c*) The transfer function $\mathbf{c}(u)$. (*d*) The field pattern $F_x(l, 0)$ (normalized Fourier transform of $g(x)$. (*e*) The effective area $\mathbf{A}(l, 0)$.

l coordinate or $\geqslant 1/d$ radians in angular measure. It follows that the multiple responses do not confuse measurements of an isolated radio source whose angular size is less than $1/d$ radians, since radiation is not received from more than one part of the source at any one time. The simple grating is used for observations of strong isolated sources such as the Sun. The multiple responses are often very useful, since the same source may be observed many times as it passes through consecutive grating beams without any readjustment of the equipment.

Steering the reception pattern of a grating antenna. The grating radiotelescope usually differs from an optical grating in that the individual units can be pointed towards different parts of the sky. This will not change the positions of the grating maxima (6.7),

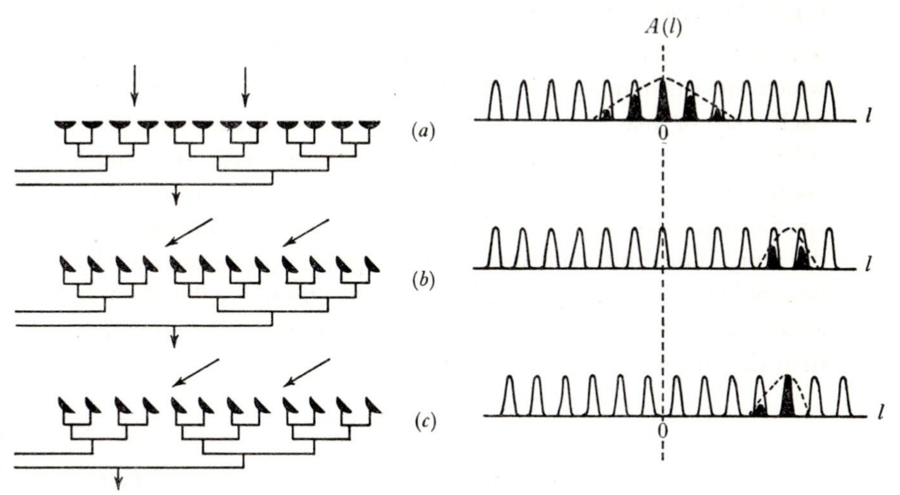

Figure 6.2. Steering the response of a grating telescope (*a*) Reception in a direction normal to the line of the grating. The arrangement of the paraboloids and transmission lines (left) and the effective area of the grating (solid black areas, right). (*b*) The paraboloids have been pointed away from the normal. The grating lobes have not moved but the region of high effective area has moved to the right to the maximum response of the paraboloids. (*c*) Electrical phase adjustment of the grating has moved the grating lobes and has centred one of these on the maximum response of the individual paraboloids.

since the quadratic factor in (6.6) depends only upon the array grading g_i, which does not change when the individual units are steered. It will, however, change the 'envelope' $N . A_e(l, m)$ so that it becomes centred on some other direction (l_0, m_0) as Figure 6.2*b* shows. The units are usually steered mechanically; it should be remembered that a large shift l_0, m_0 will change the shape of the envelope when expressed in the 'projected' l, m coordinates. The grating maxima can be shifted electrically by introducing a progressive phase shift in the array grading. A phase change $2\pi x_i l_0$ radians applied to each unit (i) at x_i, which corresponds to a change in the array grading from g_i to $g_i \exp\{j2\pi x_i l_0\}$, will shift the train of grating maxima under the envelope so that one maximum falls in the direction $l = l_0$ (see Figure 6.2*c* and also section 4.12). Some further problems concerned with the steering of the response of a grating telescope will be discussed in section 6.32.

6.1.2. *The compound grating antenna*

This antenna (compound interferometer) is a correlation telescope which consists of one grating antenna and one other antenna which itself may or may not be a grating.[602] In such an instrument (Plates 10, 11, facing p. 96) it is possible to make available a continuous spectrum of spacings (with the exception of the zero spacing) up to the maximum allowed by the extreme dimensions of the array. The multiple responses of the single grating can thereby be avoided. If A_1 and A_2 (both functions of l and m) are the effective areas of the two component parts of the compound interferometer, then the cosine and sine effective areas are given by

$$A_c = 2(A_1 A_2)^{\frac{1}{2}} \cos \psi, \tag{5.12}$$

$$A_s = 2(A_1 A_2)^{\frac{1}{2}} \sin \psi. \tag{5.15}$$

The Fourier transform of an *even* function (characterized by $g(x,y) = g^*(-x, -y)$) has a zero imaginary component (Appendix 2). The grading of an electrically steered antenna which is symmetrical about the x, y origin is an even function and it follows that the field pattern has constant $(0, \pi)$ phase in all directions. The phase difference ψ between the outputs from two electrically steered symmetrical antennas will, therefore, occur only because of the difference τ sec in the time of arrival of the waves at the centres of the two antennas. For a source in the direction l the difference in path is $d'l$ wavelength, where d' is the distance along the x-axis between the two centres of symmetry (Figure 6.3). Hence

$$\psi = 2\pi d'l \quad \text{(radians)}. \tag{6.8}$$

Combining a grating with a single antenna. If the grating has a uniform array grading then we can substitute its effective area

(6.6) for $A_1(l,m)$ in (5.12) and (5.15). With the value of ψ from (6.8) we get

$$\left.\begin{aligned} A_c(l,m) &= 2\{N\,.\,A_e(l,m)\,.\,A_2(l,m)\}^{\frac{1}{2}}\,.\,\frac{\sin NU}{N\sin U}\,.\cos\frac{2d'}{d}\,\mathrm{U},\\ A_s(l,m) &= 2\{N\,.\,A_e(l,m)\,.\,A_2(l,m)\}^{\frac{1}{2}}\,.\,\frac{\sin NU}{N\sin U}\,.\sin\frac{2d'}{d}\,U,\end{aligned}\right\} \quad (6.9)$$

where $U = \pi d \times l$ and d (wavelengths) is the distance between adjacent units in the grating.

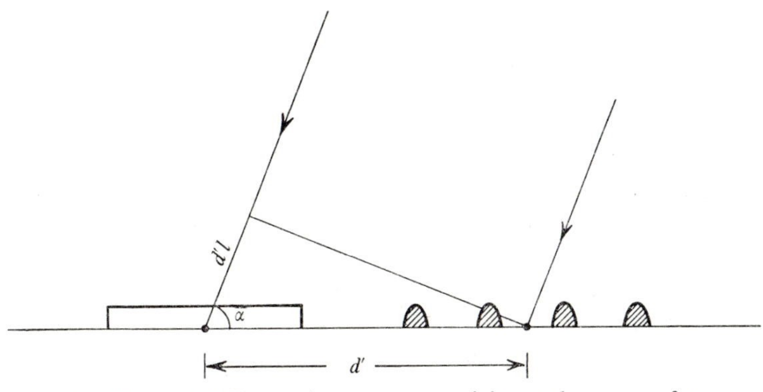

Figure 6.3. The path difference between rays arriving at the centre of symmetry of the two parts of a compound grating antenna.

The compound interferometer in its original form consisted of a single antenna placed at one end of a grating so that the distance d' between the centre of the single antenna and the grating was about $Nd/2$ wavelengths (Figure 6.6). This arrangement, although often the best, as we shall demonstrate later, has serious disadvantages, particularly for the measurement of polarization. Two other arrangements are sometimes used to overcome these. The first is a symmetrical system in which the single antenna is placed at the centre of the grating ($d' = 0$). The second is the one-dimensional analogue of the grating T antenna (section 6.2.2) and has several outstandingly useful features. In this, the single antenna is placed at a distance $d' = d/4$ from the centre of the grating in line with the elements of the grating. If we substitute these values of d' in (6.9) we find:

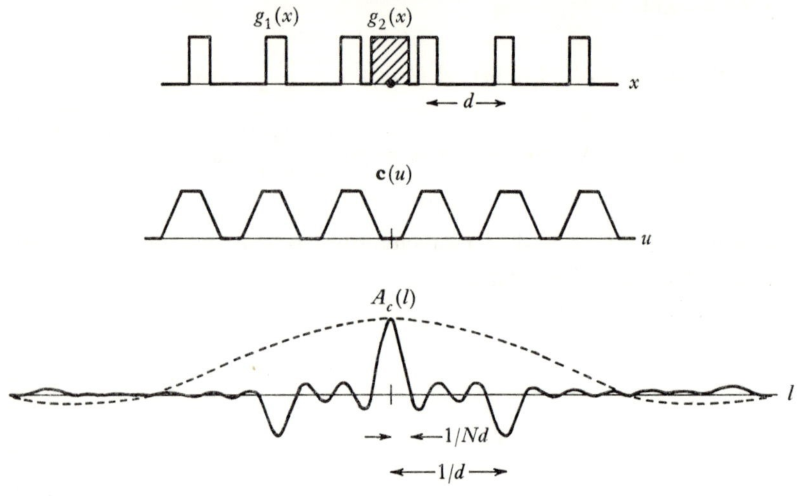

Figure 6.4. A symmetrical compound grating antenna: its arrangement, grading, cosine transfer function and cosine effective area.

(*a*) $d' = 0$ (Figure 6.4)

$$\left.\begin{aligned} A_c(l,m) &= A_0(l,m)\frac{\sin NU}{N\sin U}, \\ A_s(l,m) &= 0, \end{aligned}\right\} \tag{6.10}$$

where $$A_0(l,m) = 2\{N \,.\, A_e(l,m)\, A_2(l,m)\}^{\frac{1}{2}}$$
and when

(*b*) $d' = d/4$ (Figure 6.5)

$$\left.\begin{aligned} A_c(l,m) &= A_0(l,m)\frac{\sin NU}{2N\sin\frac{1}{2}U}, \\ A_s(l,m) &= A_0(l,m)\frac{\sin NU}{2N\cos\frac{1}{2}U} \end{aligned}\right\} \tag{6.11}$$

and when

(*c*) $d' = Nd/2$ (Figure 6.6)

$$\left.\begin{aligned} A_c(l,m) &= A_0(l,m)\frac{\sin 2NU}{2N\sin U}, \\ A_s(l,m) &= A_0(l,m)\frac{\sin^2 NU}{N\sin U}. \end{aligned}\right\} \tag{6.12}$$

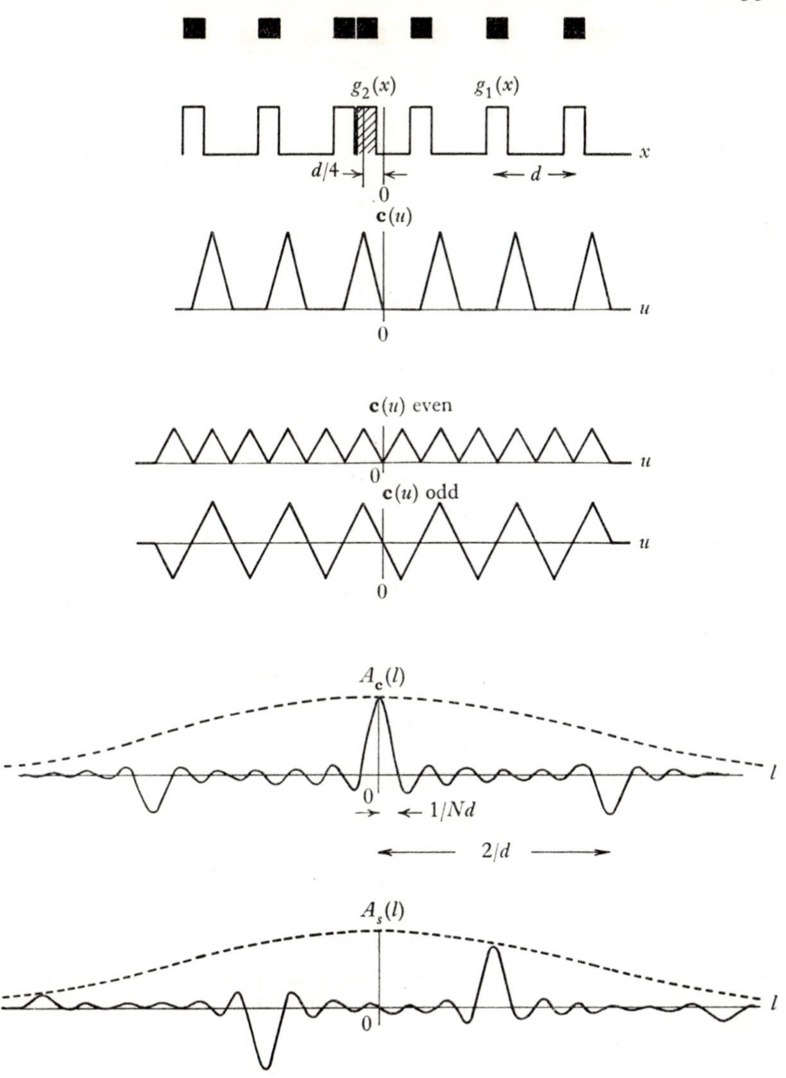

Figure 6.5. A quasi-symmetrical compound grating antenna: its arrangement, grading, transfer function with even and odd components, and its cosine and sine effective areas.

It is profitable to study (6.10), (6.11) and (6.12). The cosine effective areas for each of the three arrangements have the same general form

$$A_c(l, m) = A_0(l, m) \cdot \frac{N'' \sin N'U}{N' \sin N''U}. \tag{6.13}$$

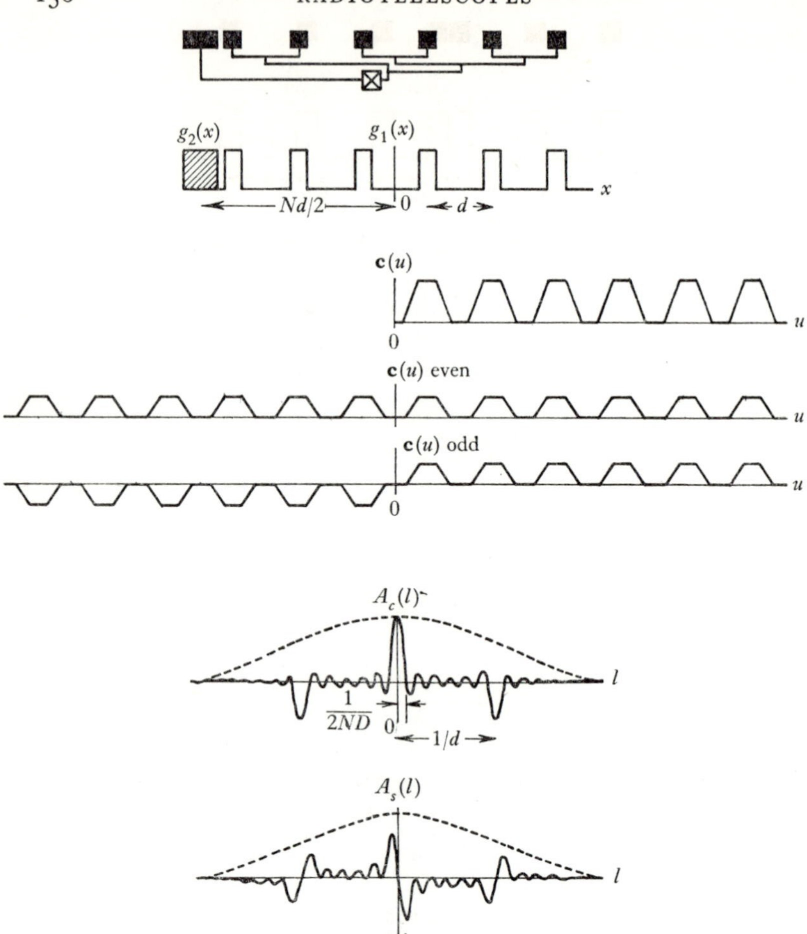

Figure 6.6. The asymmetrical compound grating antenna: arrangement, grading, transfer function, even and odd components, and cosine and sine effective areas.

If $A_0(l,m)$ is neglected and $N' \gg N''$, (6.13) takes the form, approximately, of $\sin N'U/N'U$ near $U = 0$, but, unlike that function, it repeats itself at intervals $\delta(N''U) = 2\pi$ with maxima (positive and negative) at $N''U = k\pi$, where $k = 0, 1, 2$, etc. The cosine effective areas of (6.10) and (6.11) are similar in form near the maxima (Figures 6.4 and 6.5), whereas that of (6.12) varies more rapidly with U (Figure 6.6). The cosine effective area of (6.10) is similar in

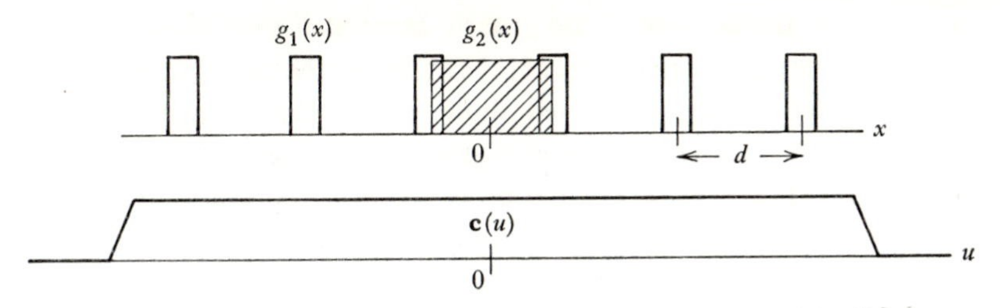

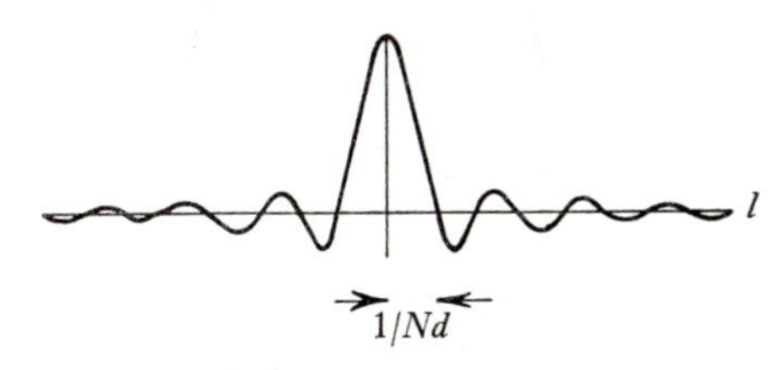

Figure 6.7. A symmetrical compound grating telescope in which the single antenna has a width d.

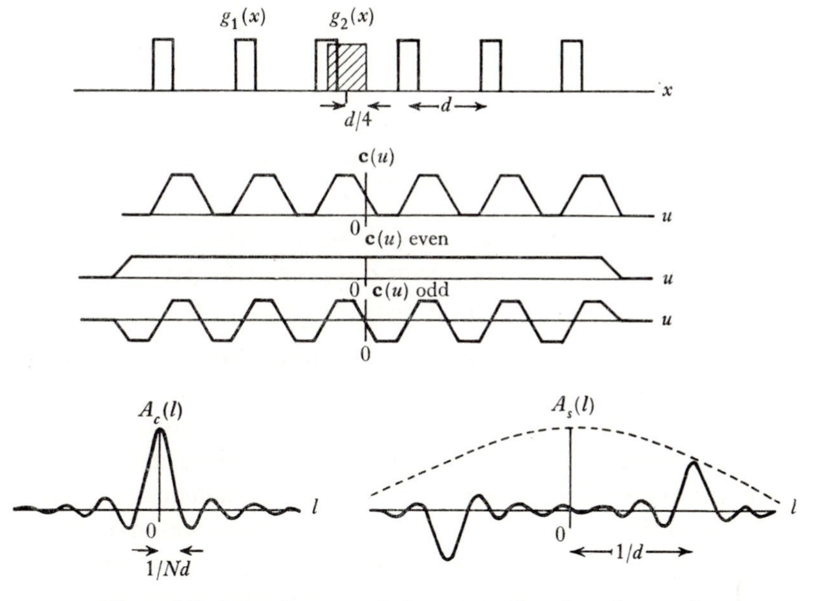

Figure 6.8. A quasi-symmetrical compound grating telescope in which the single antenna has a width $d/2$.

form to the *field pattern* of the grating antenna if used alone, while the cosine effective area of (6.12) is similar to the field pattern of a grating twice as long, also used alone. Hence the arrangement (*c*) is superior in resolving power to the arrangements (*a*) and (*b*). These two, however, are superior to (*c*) in some other respects: (*a*) has the outstanding characteristic (6.10) that its sine effective area is zero. The advantage of this is seen if we take into account the possibility of a phase error ϕ when we combine the signals from the

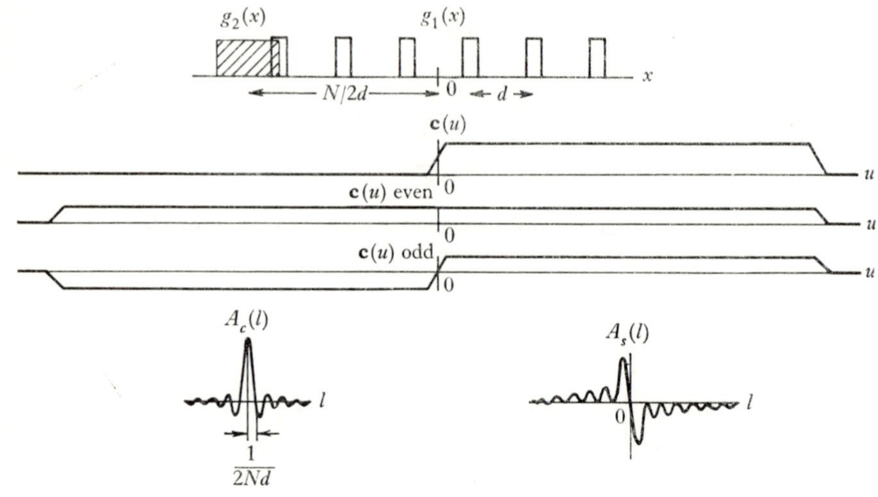

Figure 6.9. An asymmetrical compound grating in which the single antenna has a width *d*.

two parts of the system. We have shown in section 5.23 that the effective area, when the phase error ϕ is present, is given by

$$A_\phi(l, m) = A_c(l, m)\cos\phi + A_s(l, m)\sin\phi. \qquad (5.17)$$

The second term represents a distortion of the beam pattern produced by the phase error ϕ. With the antenna (*a*) $A_s(l, m) = 0$; hence this type of beam-pattern distortion does not occur. The importance of this is increased when the single antenna is of a different type from the units of the grating. Then the different antennas may not have the same relative phase centres in all directions within the field of observation. The error phase angle ϕ will then vary over the field and, if $A_s(l, m)$ is not zero, the image may be distorted in a way that cannot be corrected simply. There is a further advantage of arrangement (*a*) which should be mentioned: Polarization of the incoming radiation from a source can be detected and measured by having the feed antennas of the grating turned through $\pi/2$ compared with the feed of the single antenna. Only polarized radiation will then produce a correlated component in the two parts of the system. When the centres of the two parts of the interferometer coincide, as they do in system (*a*), the type of polarized radiation accepted by the antenna is constant over the field of observation. On the other hand, with widely separated centres as in arrangement (*c*) the polarization accepted varies over the response region of the telescope and introduces difficulties in interpreting the observations. Hence, for this type of observation arrangement (*a*) is greatly superior to arrangement (*c*).

Arrangement (*b*) has the interesting features (6.11) that (i) it repeats itself at intervals that are twice as large as with the other two arrangements, and (ii) since the centres of the two systems almost coincide, $A_g(l, m)$ is almost zero in the neighbourhood of the maxima of $A_c(l, m)$ (but not half-way between maxima). Further, the polarization accepted by the antenna is nearly constant over the individual beams of the telescope, though not over the whole field of observation.

Designing for a single-response beam pattern. The most significant feature of a compound grating telescope compared with the simple grating lies in the possibility of eliminating the multiple responses of the simple grating which limit its usefulness to observation of strong isolated sources.

What is required is that the single antenna should be of such a length and placed in such a position relative to the grating that all element spacings will be available out to the maximum, with no gaps in the transfer function. It is clear from Figure 6.9 that a uniformly graded line antenna of length equal to the spacing between adjacent elements of the grating will fulfil this requirement if it is placed close to one of the grating elements.

For such a line antenna the effective area $A_2(l, m)$ was given in (2.40) and this can be expressed as

$$A_2(l, m) = A_{2\max}\left\{\frac{\sin U}{U}\right\}^2, \tag{6.14}$$

where $U = \pi d \times l$. If we substitute this in (6.9) we find

$$\left.\begin{aligned} A_c(l,m) &= 2\{NA_e(l,m)\,A_{2\max}\}^{\frac{1}{2}}\frac{\sin NU}{NU}\,.\cos\frac{2d'}{d}\,.\,U, \\ A_s(l,m) &= 2\{NA_e(l,m)\,A_{2\max}\}^{\frac{1}{2}}\frac{\sin NU}{NU}\,.\sin\frac{2d'}{d}\,.\,U \end{aligned}\right\} \qquad (6.15)$$

and the rapidly-varying factor of the equation has been reduced to $\sin NU/NU$, which is not periodic but has a single maximum. We may note that in the arrangement (*b*), (6.11), it is necessary only to make the length of the single antenna equal to $d/2$ instead of d in order to accomplish this. Hence, we have the possibility of making the single antenna roughly equal in size to the units of the grating.

In practice, the single antenna is frequently a parabolic reflector or some antenna with non-uniform grading. The physical size of the aperture of the single antenna must then be greater than d (or, in the arrangement (*b*), $d/2$). With a standard parabolic reflector an aperture equal to $1{\cdot}5d$ (or, in arrangement (*b*), $0{\cdot}75d$) would be just sufficient. The closest subsidiary maximum would then be near the first zero of the paraboloid response (Table 3.1). We shall now show that in an image-forming, or electrically steered, system the size of the antenna has to be further increased.

Steering the reception pattern of a compound grating antenna. The pattern of the compound grating antenna can be steered in the same way as that for the simple grating. The individual antenna units may be steered mechanically to point in some direction (l_0, m_0). This will shift the envelope so that its centre lies in this direction. If the antennas have all been rotated about points on the same line so that their relative phasing is not changed, the positions of the grating maxima $l = \pm k/d$ ($k = 0, 1, 2, 3, \ldots$) remain unchanged. The most prominent lobe within the envelope will, therefore, be that for which k/d is closest to l_0. The train of grating maxima can be shifted under the envelope by phasing the individual array units. A phase change $2\pi x_i l_0$ radians applied to each unit i at x_i in both the component antennas or arrays will cause the central maximum to be shifted to the direction $l = l_0$.

The narrow beam can clearly be made to scan the part of the sky that is within the envelope by means of electrical steering alone. The electrical steering will, however, shift the whole train of grating responses under the envelope and a previously suppressed subsidiary lobe may now move in from the opposite edge of the envelope. In order to avoid this we must reduce the angular width of the envelope even further. For an image-forming telescope of this type the single telescope must be made twice as large as is required for a central beam reception only. With a somewhat smaller paraboloid the beam could still be steered undisturbed between the half-power directions of the envelope.

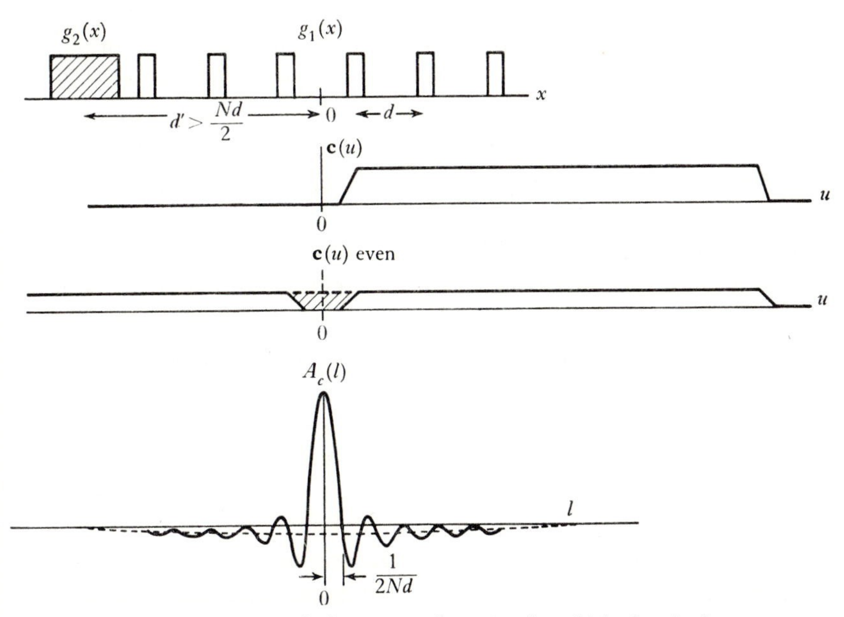

Figure 6.10. An asymmetrical compound grating in which the single antenna has been separated from the grating. The gap in $\mathbf{c}(u)$ produces a shallow negative depression in $A_c(l)$.

If, for example, we take elements of a grating to be $d/2$ wavelengths in diameter, then for suppression of subsidiary maxima over the whole image field the single antenna, presumed a paraboloid, should be three times or six times the diameter of the paraboloids of the grating, depending on whether the arrangement of Figures 6.5, 6.4, or 6.6 is used. This difference in size has the practical disadvantage that it is difficult to arrange all antennas to have their axes on the same line. When this cannot be done the processing of signals is further complicated. The second difficulty lies in the possibility of the phase centres of the paraboloids of greatly differing size shifting relative to each other for observations in different directions within the image field.

The problem of the zero spacing component. The idealized single-response compound interferometers shown in Figures 6.7, 6.8 and 6.9 must be modified in practice, since the large antenna must not overlap with any of the grating units. A simple modification of the arrangement 6.9 is shown in Figure 6.10. The large antenna is here placed further from the end of the array. The shortest spacings are now missing and the transfer function has the gap at the origin, which is characteristic of correlation telescopes with separate non-interacting antennas (sections 5.32 and 5.33). The transfer function can be described as the difference between two functions, one having a constant value out to the maximum spacing and one representing the missing portion at the centre. Hence, the effective area can be described as an ideal single-beam pattern minus a wide low-amplitude pattern which is the Fourier transform of the missing central portion. If, in (6.15), we set $d' = (N+n)\,d/2$ instead of the ideal $Nd/2$, then A_c can be written in a form which shows it as the difference between these two patterns:

$$A_c(l,m) = 2\{NA_e(l,m)\,A_{2\max}\}^{\frac{1}{2}}\cdot\left[C_1\frac{\sin(2N+n)\,U}{(2N+n)\,U} - C_2\frac{\sin nU}{nU}\right], \tag{6.16}$$

where $\quad C_1 = (2N+n)/2N \quad$ and $\quad C_2 = n/2N.$

The wide and shallow sidelobe surrounding the main beam is a common feature of single-response correlation telescopes. It results because the different antennas cannot overlap ($\mathbf{c}(0,0) = 0$) and it ensures that the reception pattern obeys 5.30 and 5.31, which states that the all-sky integral of a correlation effective area must be zero. This feature limits its usefulness. One way of overcoming it is to add to the output of the receiver a signal component which represents the missing spatial components $\mathbf{c}(u)$ near $u = 0$. These components can be supplied from a separate total power antenna.

More complex types of compound gratings. The simplest type of compound grating telescope, which we have been discussing, is not the most convenient instrument to build if very high resolving power is needed. The reason for this is that the maximum spacing between grating elements is limited to a fraction of the diameter of the single antenna; hence, if the total aperture of the telescope is to be very great, either the single antenna must be large or else the grating elements must be very numerous. This difficulty can be overcome if the single antenna is replaced by a second grating which has an element spacing equal to the length of the first grating.(603–6) One form of double grating is shown in Figure 6.11*a*. The smaller grating has N elements with spacing d wavelengths and the larger

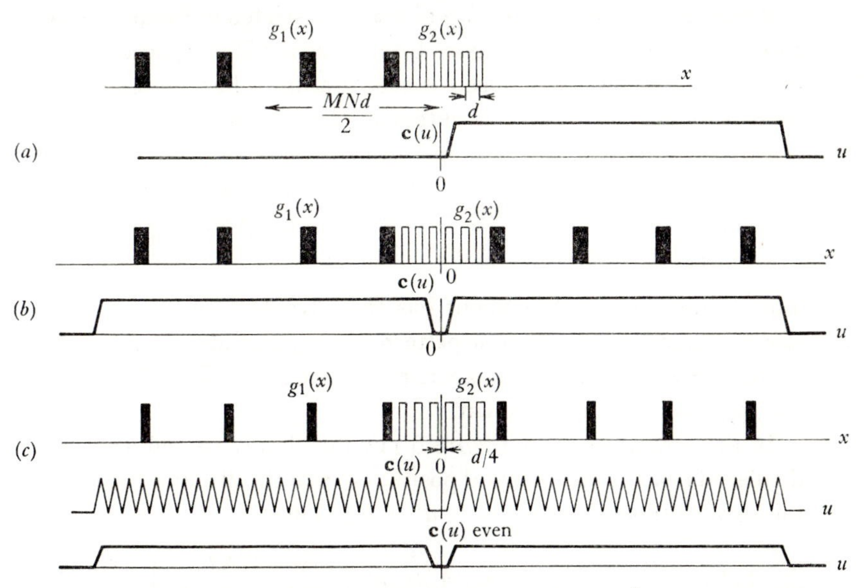

Figure 6.11. Compound grating telescopes composed of two gratings with wide and narrow spacings. (*a*) Asymmetrical form. (*b*) Symmetrical. (*c*) Quasi-symmetrical.

one has M elements with spacings Nd. If we substitute for $A_e(l, m)$ in 6.9 the effective area of the second grating and for d' the distance $MNd/2$ between the centre of the two gratings, then

$$\left.\begin{aligned} A_c(l,m) &= 2\{MN\,A_{e1}(l,m)\,A_{e2}(l,m)\}^{\frac{1}{2}} \\ &\qquad\times\frac{\sin NU}{N\sin U}\cdot\frac{\sin MNU}{M\sin NU}\cdot\cos MNU \\ &= 2\{MN\,A_{e1}(l,m)\,A_{e2}(l,m)\}^{\frac{1}{2}}\frac{\sin 2MNU}{2MN\sin U}, \\ A_s(l,m) &= 2\{MN\,A_{e1}(l,m)\,A_{e2}(l,m)\}^{\frac{1}{2}}\frac{\sin^2 MNU}{MN\sin U}, \end{aligned}\right\} \quad (6.17)$$

A_{e1} and A_{e2} are the effective areas of the individual elements in the two gratings. The reception pattern of this double grating compound interferometer, which has a total of only $(M+N)$ antennas, is, apart from the shape of the envelope, the same as that of a simple form of compound interferometer which has a total of $(MN+1)$ antennas. This double grating has the same features as the simpler compound grating of Figure 6.9. Double gratings that are analogous to those shown in Figures 6.7 and 6.8 are illustrated in Figure 6.11 *b* and *c*. The first of these (Figure 6.11 *b*) has a widely spaced array placed at each end of the closely spaced grating. The widely spaced units form an array with $2M$ antennas. The distance between the centres of symmetry is $d' = 0$ and we get

$$\left.\begin{aligned} A_c(l,m) &= 2\{2MNA_{e1}(l,m)\,A_{e2}(l,m)\}^{\frac{1}{2}}\,\frac{\sin 2MNU}{2MN\sin U}, \\ A_s(l,m) &= 0. \end{aligned}\right\} \qquad (6.18)$$

This symmetrical arrangement has the advantage of a zero sine effective area (and an envelope pattern with a $\sqrt{2}$ greater amplitude), but the reception pattern is, apart from this, identical with that of the asymmetrical arrangement.

In the arrangement of Figure 6.11 *c* the closely spaced array is displaced from its symmetrical position so that the distance between the two centres of symmetry is $d' = d/4$. Then

$$\left.\begin{aligned} A_c(l,m) &= 2\{2MN\,A_{e1}(l,m)\,A_{e2}(l,m)\}^{\frac{1}{2}}\,\frac{\sin 2MNU}{2MN\,.\,2\sin\frac{1}{2}U}, \\ A_s(l,m) &= 2\{2MNA_{e1}(l,m)\,A_{e2}(l,m)\}^{\frac{1}{2}}\,\frac{\sin 2MNU}{2MN\,.\,2\cos\frac{1}{2}U}. \end{aligned}\right\} \qquad (6.19)$$

This nearly symmetrical arrangement gives the telescope similar properties to those of the simpler nearly symmetrical arrangement which we discussed earlier (Fig. 6.8).

6.1.3. *Beam shaping and restoring*

Non-uniform array gradings. The grating antennas in the compound interferometers discussed so far have had uniform array gradings. The transfer functions of these telescopes have, therefore, a constant amplitude envelope out to the maximum available spacing. This means that the shape of the main (cosine) beams has been given by functions of the general type sin U/U. Such patterns have rather prominent sidelobes which may be undesirable in many circumstances. The envelope of the transfer function must be graded or tapered if its Fourier transform, the telescope effective area, is to have lower sidelobes. This can sometimes be achieved by tapering the array grading of the grating component: an example is given in Figure 6.12. The effective area cannot be calculated in a

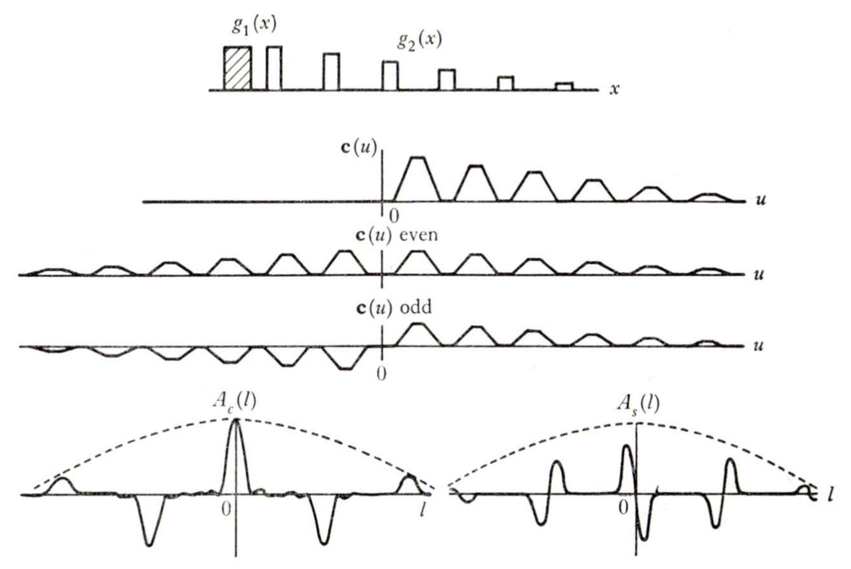

Figure 6.12. A tapered compound grating telescope.

simple way from equations of the type (5.12) and (5.15), since the grating will no longer be an electrically symmetrical antenna and the phase difference ψ does not simply depend on the distance d' between the mechanical centres of symmetry. For this, as indeed for most other calculations of effective area, the use of the transfer function

$$\mathbf{c}(u) = \text{const.}\, g_1(x) \star g_2^*(x) \tag{5.42}$$

usually provides a simpler solution. Rapid approximate calculations of $\mathbf{c}(u)$ for a one-dimensional correlation telescope can be made by representing the gradings $g_1(x)$ and $g_2(x)$ graphically and sliding one over the other, as Figure 5.3 illustrates. The transfer functions shown in the figures of this chapter can all be reproduced in this simple way. The complex effective area $\mathbf{A} = A_c + jA_s$ is then equal to the Fourier transform of $\mathbf{c}(u)$

$$\mathbf{A}(l) \mathscr{F} \mathbf{c}(u) \qquad (5.41)$$

and it follows that A_c and A_s are the transforms of the even and odd parts respectively of $\mathbf{c}(u)$ (Appendix 2). The constant in (5.42) is that which makes

$$|\mathbf{A}|_{\max} = 2(A_{1\max} A_{2\max})^{\frac{1}{2}}. \qquad (5.20)$$

In practice we calculate directly the Fourier transform of $g_1 \star g_2^*$ and then normalize $\mathbf{A}(l)$ so that this relation is satisfied.

'Restoring' after the measurements have been made. It is often not possible to achieve a suitable reception pattern simply by manipulating the gradings of the component antennas. The preferred pattern shape corresponds to a particular shape of the transfer function, since $\mathbf{A} \mathscr{F} \mathbf{c}$. For a two-grating interferometer, for example, there are $M.N$ maxima in the transfer function which must have the particular values which correspond to the preferred shape of the beam, whereas there are only $(M+N)$ array gradings that can be adjusted. Clearly this will be possible only in special cases and, as we have seen, the evenly weighted transfer function which gives a $\sin U/U$ type pattern is one of these special cases.

Instead of adjusting the telescope gradings for a particular preferred beam shape we can adjust the measurements of a radio source so that they are 'restored' to what they would be if they had been measured with the preferred beam. In section 5.41 we have shown that the brightness temperature distribution over the sky as measured by a radiotelescope is the true brightness distribution smoothed by the antenna response. Leaving out the indices of l_0 and m_0 from (5.52) we have

$$\mathbf{T}_a(l, m) = \lambda^{-2}[T_B(l, m) \star \mathbf{A}(l, m)]. \qquad (5.52)$$

$\mathbf{T}_a(l, m)$ is the measured antenna (or correlation) temperature when the beam is steered to the direction (l, m). The same relation could also be expressed in terms of the telescope transfer function:

$$\mathbf{T}_a(l, m) \mathscr{F} \lambda^{-2}[\mathbf{t}(u, v) . \mathbf{c}(u, v)], \qquad (5.54)$$

where $\mathbf{t}(u, v)$ is the Fourier transform of the true brightness distribution $T_B(l, m)$. We want to restore the actual measurements $\mathbf{T}_a(l, m)$ to the form $\mathbf{T}_a^r(l, m)$ which they would have if they had been made with a different beam $\mathbf{A}^r(l, m)$ that corresponds to a different transfer function $\mathbf{c}^r(u, v)$—i.e. $\mathbf{A}^r(l, m) \mathscr{F} \mathbf{c}^r(l, m)$. Now

$$\mathbf{T}_a^r(l, m) \mathscr{F} \lambda^{-2}[\mathbf{t}(u, v) \,.\, \mathbf{c}^r(u, v)] \tag{6.20}$$

which can be written

$$\mathbf{T}_a^r(l, m) \mathscr{F} \lambda^{-2}[\mathbf{t}(u, v) \,.\, \mathbf{c}(u, v)] \,.\, r(u, v), \tag{6.21}$$

where
$$r(u, v) = \mathbf{c}^r(u, v)/\mathbf{c}(u, v) \tag{6.22}$$

wherever $\mathbf{c}(u, v) \neq 0$. The value of r over u, v regions where $\mathbf{c} = 0$ will clearly make no difference to the Fourier relation 6.21. $\mathbf{T}_a^r$ is here expressed as the Fourier transform of a product of two factors; consequently, it can also be written as the convolution integral of the individual transforms of the two factors (convolution theorem—see Appendix 2). Equation 5.54 shows that the first factor has the inverse transform $\mathbf{T}_a(l, m)$.

Hence
$$\mathbf{T}_a^r(l, m) = \mathbf{T}_a(l, m) \star R(l, m), \tag{6.23}$$

where $R(l, m)$ is the transform of $r(u, v)$. We see that the brightness distribution as measured with the telescope can be restored to what it would be if the telescope had had a different preferred reception pattern $\mathbf{A}^r(l, m)$ by smoothing the measured distribution with a *restoring function* $R(l, m)$. The form of $r(u, v)$ over the regions in the (u, v) plane where $\mathbf{c}(u, v) = 0$ makes no difference to $\mathbf{T}_a^r(l, m)$ as calculated from (6.21). This means that the restoring procedure may be used to *change the weighting of* the available antenna spacings, but it cannot compensate for those completely missing. Figure 6.13 illustrates a simple restoration applied to a line aperture.

The measurements of any particular radio source, clearly, can be presented in different versions depending upon how the restored transfer function $\mathbf{c}^r(u, v)$ has been chosen. The uniform weighting, $\mathbf{c}^r = 1$ over the available u, v region, gives a picture of the radio source which is known as the *principal solution*.[503] It has the unique property of giving the best root mean square fit to the true brightness distribution over the source, but the 'wriggles' caused by the rather prominent sidelobes in the corresponding sin U/U type telescope pattern often give the principal solution a rather artificial appearance. More tapered transfer functions giving somewhat wider beams but with smaller sidelobes are usually preferred.

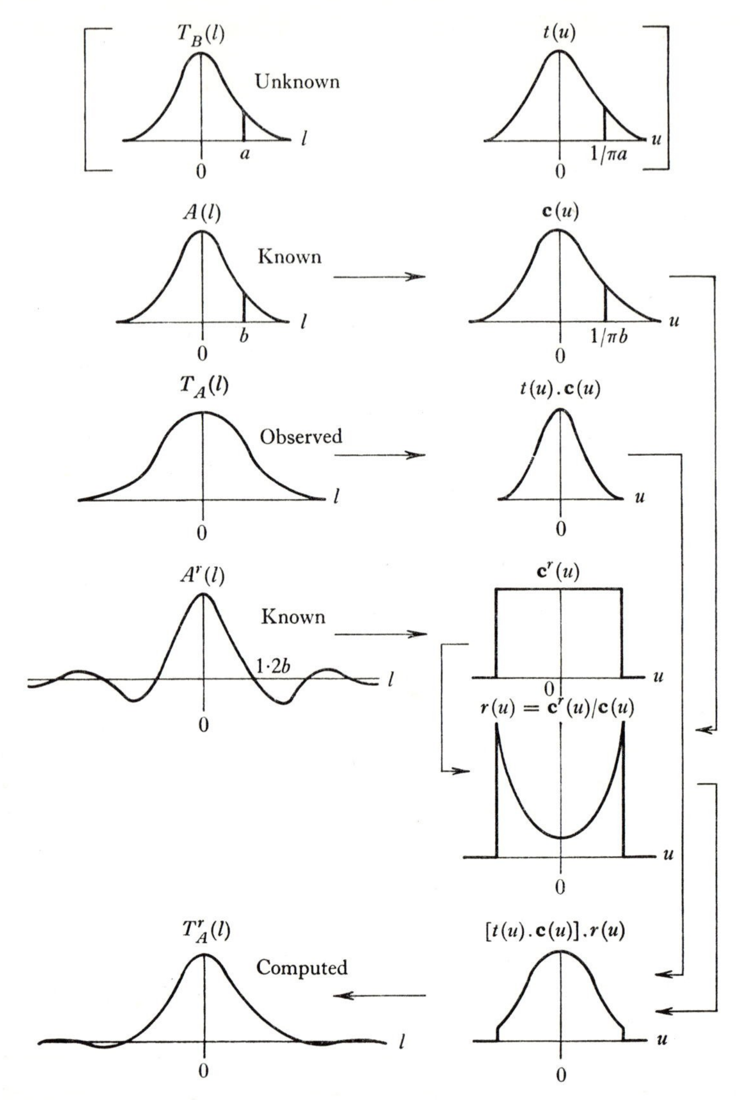

Figure 6.13. The observations $T_A(l)$ of a brightness distribution $T_B(l)$ by means of an antenna characteristic $A(l)$ are restored to $T_A^r(l)$, which is what they would have been if the observations had been made with a preferred antenna characteristic $A^r(l)$.

6.2. Two-dimensional unfilled apertures

There are many possible arrangements by means of which a two-dimensional coverage of antenna spatial components (spacings) may be obtained from unfilled apertures. These include the earliest and most used form, the cross antenna, with its variants, the grating cross, the T and grating T, and circular arrangements such as a thin ring of antennas.

6.2.1. *The cross antenna*

This antenna was described briefly in Chapter 1 and in section 5.27. The cosine and sine effective areas of a thin uniformly graded cross of length L wavelengths is approximately

$$A_c(l,m) = |\mathbf{A}|_{\max}\frac{\sin \pi Ll}{\pi Ll}\cdot\frac{\sin \pi Lm}{\pi Lm}, \tag{5.33}$$

$$A_s(l,m) = 0, \tag{5.34}$$

where
$$|\mathbf{A}|_{\max} = 2(A_{1\max}A_{2\max})^{\frac{1}{2}}. \tag{5.20}$$

The sine effective area is zero as a result of the coincidence of the centres of symmetry of the two antennas. As we have already mentioned, this makes a correlation telescope less susceptible to effects from certain types of errors in the phase adjustment between the component antennas. The all-sky integral of (5.33) is not zero as it ought to be for a correlation telescope with separate not interacting antennas. In deriving this equation, however, the implicit assumption was made that both antennas passed through the centre of the cross without interfering with one another, which, of course, is not possible in practice. It is usually necessary to leave a gap at the centre of one of the orthogonal components of the cross. The absence of the centre section in the one antenna will produce a deep trough through the origin in the telescope transfer function (Fig. 5.4). Thus there will be a wide and shallow negative sidelobe (Fig. 5.2) in the reception pattern. The all-sky integral of this pattern will be zero. The zero spacing problem is here very similar to that for line apertures which we discussed in section 6.1.2. In practice this negative response can be partly compensated by adding some of the total-power response of the unbroken arm of the cross to the output of the correlation receiver. Alternatively, we may amplify separately the signals from the central section of the unbroken antenna and add some of this signal to that from the antenna which lacks a central section.

Equation (5.23) shows that the uniformly graded cross antenna has large sidelobe responses in directions where $l = 0$ or $m = 0$. Its power response (effective area) is similar to the *field pattern* of a uniformly graded rectangular aperture (section 2.42). It is usual to taper the grading along the cross arms to diminish these responses. One suitable grading is illustrated in Figure 4.2*a*. It is interesting to note that, if a triangular grading is used in each arm of the cross, the expression for the effective area of the cross (5.23) becomes

$$A_c(l, m) \approx |\mathbf{A}|_{\max}\left[\frac{\sin(\pi Ll/2)}{\pi Ll/2}\right]^2 \cdot \left[\frac{\sin(\pi Lm/2)}{\pi Lm/2}\right]^2 \qquad (6.24)$$

which is the same as that for a uniformly graded rectangular aperture with sides $L/2$ wavelengths long. The value of $|\mathbf{A}|_{\max}$ in the two cases is, of course, very different, being L^2 (square wavelengths) for the filled aperture and approximately $1{\cdot}5\,LW$ for the cross (W is the width of the cross antennas and $W \ll L$). The effective cosine area of the cross given by the approximate eqn (5.33) can generally be used with little error, but it should be remembered that there are several factors missing from it. One is the broad negative response produced by the missing centre of one array of the cross. The others are (i) an expression in l which is related to the width of one arm, (ii) an expression in m which depends on the width of the other arm. A third factor, which is related to the plane of polarization accepted by the antenna (see (2.36)) usually has negligible effect on A_c.

The original Mills cross antenna consisted of two long thin arrays of parallel dipoles placed above a horizontal plane reflecting screen, one array being in a north–south line and the other in an east–west line.[(502)] The east–west array was permanently connected so that the maximum response lay along the meridian circle. This meant that the instrument was a meridian transit telescope. The reception pattern of the north–south array could be steered electrically to any direction on the meridian within a restricted range of angles about the zenith. The electrical distance between each dipole and the receiver, and hence the phase of the array grading, could be set for maximum response in some particular direction by adjusting the positions of the probes which coupled the individual dipoles to the common transmission (feeder) line leading to the receiver. In Chapter 4 we mentioned the difficulties caused by the electromagnetic coupling between adjacent dipoles, and Figure 4.6 illustrates one means used to overcome these difficulties in a cross antenna. For observations near the zenith, when all the reactive loads on the transmission line are additive, compensative loading of the line is necessary to prevent the resultant load at the receiver from becoming highly reactive.

A pair of crossed parabolic cylinders was the logical development of the arrays of dipoles, since the effective area of the cross can be increased very considerably with no increase in the electrical complexity.(607–8) With the increased directivity of the arrays, however, it was necessary to introduce mechanical steering of the east–west cylindrical reflector. The north–south array could have its responses adjusted electrically, as could the original cross antenna. Since this was intended to be used only near the meridian plane, no steering in azimuth was required. Such a cross antenna is shown in Plate 12 (facing p. 96). Its operation is similar to that of the simple cross.

The features of this type of cross antenna are, first, that it has a simple structure and, secondly, that there are no discontinuities along the individual arrays, apart from the break at the centre of the east–west array, and there are no sidelobes that cannot be reduced to a low level by control of the illumination taper (grading) along the arrays. The main undesirable feature is the necessity to control individually the phase of each dipole of the north–south array. In a large cross antenna with a resolving power of one minute of arc the number of dipoles will be of the order of 10,000 and the individual adjustment of these, complicated as it is by the necessity of avoiding mutual-coupling effects, has caused some designers to seek other ways of constructing the cross antenna.

Grating crosses. The two continuous arrays of a cross may be replaced by gratings.(603, 609) If each grating consists of an array of steerable parabolic dishes, two important new features are introduced. First, the individual dishes may be steered mechanically in both hour angle and declination, so that the cross is no longer a meridian transit device. Secondly, the number of elements to be phased electrically is drastically reduced in comparison with a dipole array, and mutual coupling of adjacent elements is reduced to negligible proportions so that simple phasing methods may be used.

The main characteristic of such a grating cross, as shown in Figure 6.14*a*, is that it has multiple responses. A grating cross which

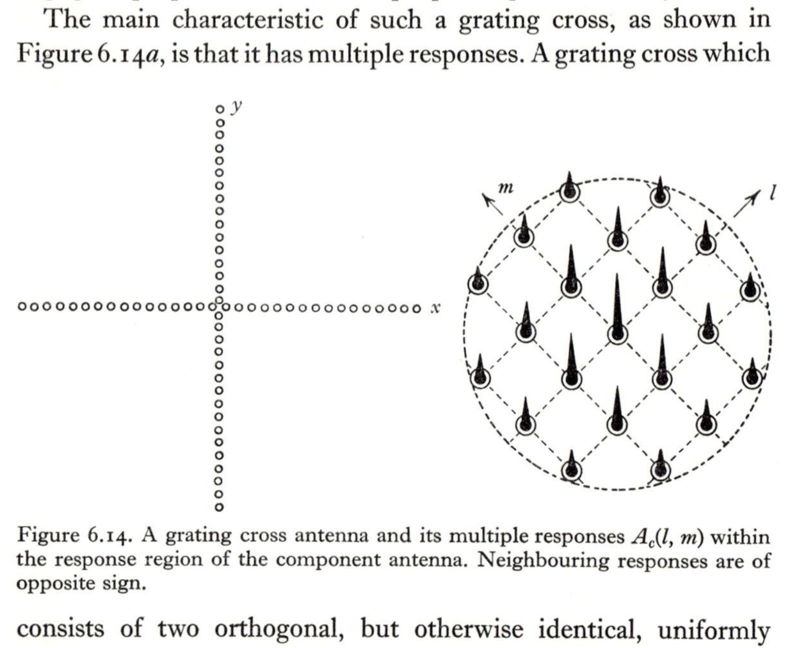

Figure 6.14. A grating cross antenna and its multiple responses $A_c(l, m)$ within the response region of the component antenna. Neighbouring responses are of opposite sign.

consists of two orthogonal, but otherwise identical, uniformly graded arrays of N units each, with coinciding centres of symmetry and no internal losses, will have the effective area

$$\left.\begin{aligned} A_c(l,m) &= 2NA_e(l,m)\,.\frac{\sin NU}{N\sin U}\cdot\frac{\sin NV}{N\sin V},\\ A_s(l,m) &= 0, \end{aligned}\right\} \qquad (6.25)$$

where $$U = \pi d\times l, \quad V = \pi d\times m$$

and d (wavelengths) is the spacing between adjacent elements in the arrays. The envelope of the grating pattern for both arrays is $NA_e(l,m)$ and they differ only in the orientation of the grating pattern under the envelope as expressed by the two last factors in A_c. The grating pattern of the cross consists of pencil beams in directions for which sin U and sin V *both* are equal to zero and the amplitudes of these maxima are given by the envelope in the relevant directions (Figure 6.14*b*).

The reception pattern of the grating cross may be steered in the same way as that of a single grating antenna. All that is needed is that the patterns of the two compound arrays be steered separately in the desired direction by the combination of mechanical and electrical steering described in section 6.1.1. The envelope does not change when expressed in angular measure, because it is mechanically steered (which means that it does change shape when expressed in the projected coordinates l and m). The grating beams, on the other hand, are electrically steered and their form is everywhere the same expressed in the l, m coordinates.

This type of grating cross is limited in its usefulness to strong sources; in fact, such crosses have been used almost exclusively for high resolution studies of the Sun. For solar observations it is necessary, of course, to design the gratings so that only one of the multiple responses can fall on the Sun at a time.

In order to make a grating cross more generally useful it is necessary to devise ways in which the multiple grating responses can be reduced to a single response. Figure 6.15 shows one way in which this has been done. The north–south array is a grating made up of small cylindrical parabolic antennas, and the east–west arm is a single (except for a break in the centre) cylindrical parabolic antenna which has its narrow dimension considerably larger than the spacing between adjacent elements of the north–south grating. This cross, in fact, suppresses the unwanted responses in the same way as, in one dimension, the compound interferometer does. In the design of the Bologna cross antenna,[409] which is of this type (Plate 13, facing p. 96), care was taken that the north–south response of the east–west antenna had no significant magnitude in any direction which would coincide with more than one of the grating responses of the north–south array. Sidelobes were reduced by feeding the large cylindrical reflector with great care so that a suitable grading was produced across the narrow dimension of the aperture. Another step taken was to eliminate the blocking of the aperture by the feed. We have seen that feed blocking in a cylindrical paraboloid can produce a significantly large and wide negative response around the main antenna response. With the grating cross this would be particularly harmful; an off-aperture feed similar to that shown in Figure 4.9 was used to overcome this.

The grating cross just described, because of its cylindrical para-

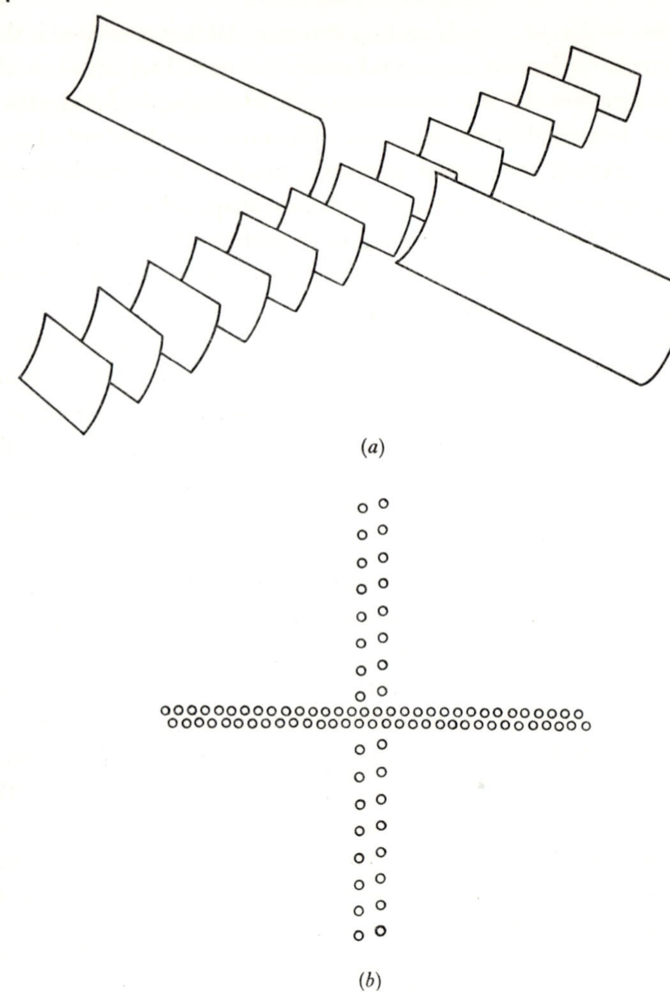

Figure 6.15. Grating cross antennas which have a single maximum in their cosine patterns. (*a*) Parabolic cylinders. (*b*) Parabolic dishes.

boloid reflecting surfaces, is necessarily a meridian transit (or near-meridian transit) instrument. The more ambitious single-response grating cross shown in Figure 6.15*b*, made up entirely from steerable parabolic dishes, was designed but not constructed.[610] The spacings between elements of the two (multiple) gratings have been so made that a strong response from one grating does not coincide with any significant response from the other grating.

6.2.2. *The T antenna*

One-quarter of a cross antenna provides redundant information, since all element spacings u, v are still present if half of one array is removed. The remaining 'T antenna' has a transfer function whose even part is similar to that of the full cross antenna; the odd part, however, is different from zero, since we are no longer dealing with a symmetrical arrangement. Consequently, the cosine effective area is similar to that of the full cross, while the sine effective area is different and non-zero.

For *the simple T antenna*, with uniformly graded component antennas and no internal losses, we get from (3.5), (5.8) and (5.13) and with $\psi = \pi\frac{1}{2}Lm$

$$\left.\begin{aligned} A_c(l,m) &= |\mathbf{A}|_{\max}\cdot\frac{\sin\pi Ll}{\pi Ll}\cdot\frac{\sin\pi\frac{1}{2}Lm}{\pi\frac{1}{2}Lm}\,.\cos\pi\tfrac{1}{2}Lm,\\ A_s(l,m) &= |\mathbf{A}|_{\max}\cdot\frac{\sin\pi Ll}{\pi Ll}\cdot\frac{\sin\pi\frac{1}{2}Lm}{\pi\frac{1}{2}Lm}\,.\sin\pi\tfrac{1}{2}Lm,\end{aligned}\right\}\qquad(6.26)$$

or

$$\left.\begin{aligned} A_c(l,m) &= |\mathbf{A}|_{\max}\cdot\frac{\sin\pi Ll}{\pi Ll}\cdot\frac{\sin\pi Lm}{\pi Lm},\\ A_s(l,m) &= |\mathbf{A}|_{\max}\cdot\frac{\sin\pi L.}{\pi Ll}\cdot\frac{\sin^2\pi\frac{1}{2}Lm}{\pi\frac{1}{2}Lm}.\end{aligned}\right\}\qquad(6.27)$$

$|\mathbf{A}|_{\max} = 2(A_{1\max}A_{2\max})^{\frac{1}{2}}$ is a factor $\sqrt{2}$ lower than that of the corresponding cross, since one of the two component antennas has been halved. If we wish, we can compensate for this by making the remaining half-antenna twice as wide.

Figure 6.16 shows the cosine and sine responses in the plane of the short arm of the T antenna. As with the asymmetrical form of the compound interferometer (Figure 6.6) the T antenna is very sensitive to phase errors in the connection of the two arms. The effect of a phase error in the cosine connection is to reduce the cosine component slightly and introduce a sine component which distorts the main beam and produces a marked increase in the side-lobe response. Equations (6.26) and (6.27) refer to a T antenna with uniformly graded components In the same way as for the cross antenna we can reduce the sidelobes arising from the $\sin U/U$ type pattern by tapering the gradings of the three half-antennas. The even part of the transfer function is clearly also in this case similar to that of the cross and results in a similar cosine pattern A_c. The sine pattern must here be calculated as the Fourier transform of the odd part of the transfer function, since one of the component antennas has an asymmetrical grading and $\psi \neq \pi\frac{1}{2}Lm$.

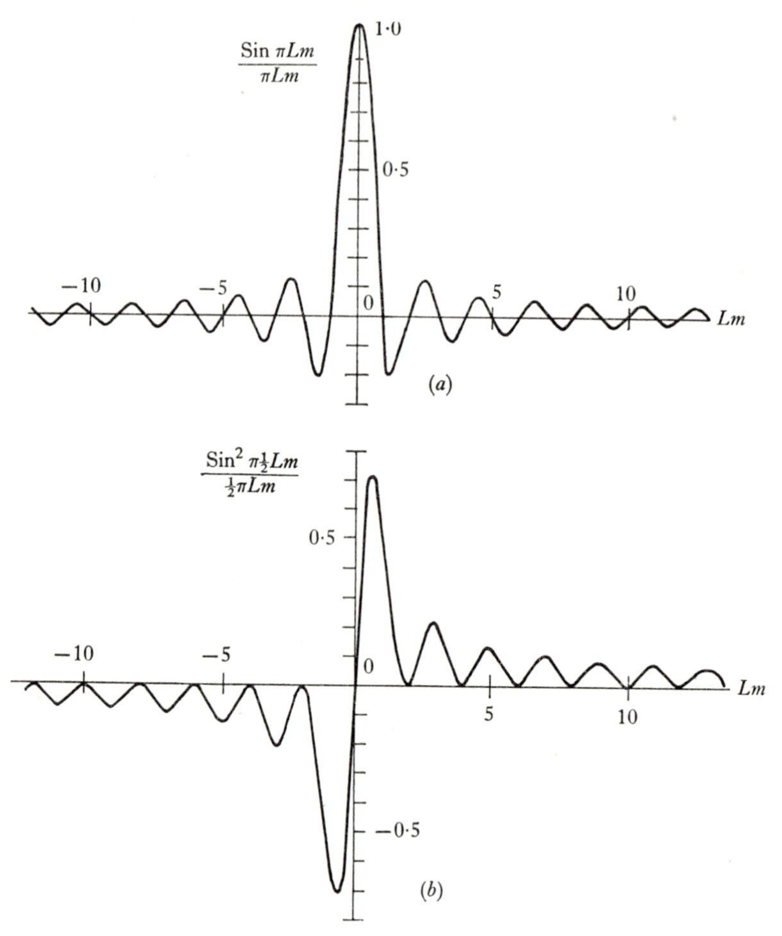

Figure 6.16. (*a*) The cosine, and (*b*) the sine effective areas of a uniformly graded *T* antenna in the plane of the short arm.

The zero spacing problem remains after half of one antenna has been removed. The various solutions and approximations discussed in dealing with the cross antenna may be applied also here and their effects on the cosine reception pattern will be much the same.

The grating T antenna was devised to overcome the difficulties involved in steering the main response of the north–south array of a cross antenna.[408] The north–south arm of a T antenna is broken into N segments (Figure 6.17) and alternate segments are transferred to the opposite side of the east–west array, as Figure 6.17 shows. All the spatial components of the antenna that are present in Figure 6.17*a* are also present in Figure 6.17*b*. The north–south

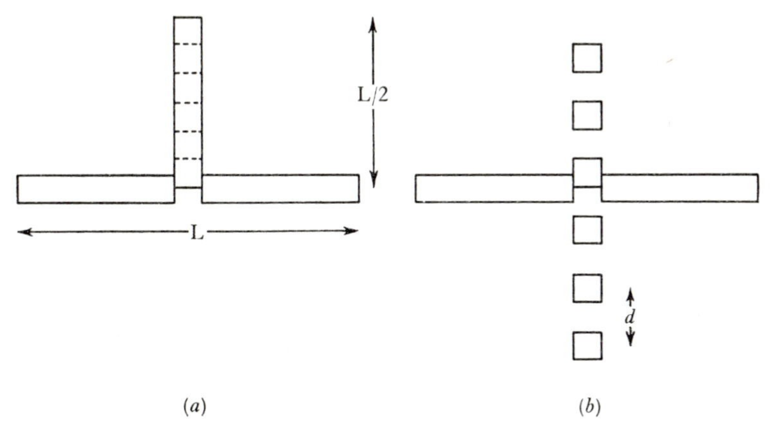

Figure 6.17. A grating T antenna. Alternate sections of the short arm of the T antenna (*a*) are transferred to the opposite side of the long arm. The characteristics of this antenna (*b*) near the main beam are similar to those of a full cross but phasing errors may produce grating sidelobes.

antenna has become a grating antenna with a uniform array grading which is offset an amount of $d' = d/4$ wavelengths from its symmetrical position with respect to the east–west antenna. The individual units are uniformly graded apertures $\frac{1}{2}d$ wavelengths wide. If, as before, we ignore the zero spacing problem and the influence on the pattern of the finite width W_1 and W_2 of the antennas, we get

$$A_c(l, m) = |\mathbf{A}|_{\max} \cdot \frac{\sin \pi Ll}{\pi Ll} \cdot \frac{\sin \pi Ndm}{N \sin \pi dm} \cdot \frac{\sin \pi \frac{1}{2}dm}{\pi \frac{1}{2}dm} \,.\, \cos 2\pi \tfrac{1}{4}dm. \tag{6.28}$$

The first trigonometric factor is due to the east–west antenna, the second and third represent the grating pattern and the pattern of the individual units of the north–south antenna, and the last factor is the cosine interferometer factor $\cos \psi$ for the $d/4$ wavelength north–south spacing between the antenna centres of symmetry. The expression for A_s differs only in that $\cos\psi$ is replaced by $\sin\psi$. Now $Nd = L$ and 6.28 gives

$$\left.\begin{aligned} A_c(l,m) &= |\mathbf{A}|_{\max}\frac{\sin\pi Ll}{\pi Ll}\cdot\frac{\sin\pi Lm}{\pi Lm}, \\ A_s(l,m) &= |\mathbf{A}|_{\max}\frac{\sin\pi Ll}{\pi Ll}\cdot\frac{\sin N\pi dm}{N\sin\pi dm}\cdot\frac{\sin^2\pi\frac{1}{2}dm}{\pi\frac{1}{2}dm} \\ &= |\mathbf{A}|_{\max}\frac{\sin\pi Ll}{\pi Ll}\cdot\frac{\sin\pi Lm}{\pi Lm}\cdot\tan(\pi\tfrac{1}{2}dm). \end{aligned}\right\} \qquad (6.29)$$

$A_c(l,m)$ is, as we expected, identical with that of the simple T antenna (6.27). The sine effective area, however, is different and much smaller than A_c in the vicinity of the main beam at $m = 0$ because of the slowly varying factor $\tan\pi\frac{1}{2}dm$. An error in the relative phasing of the two component antennas, therefore, will not produce a distorted main beam and close sidelobes but will, instead, give rise to some distant sidelobes in the directions of those grating maxima which are suppressed by the $d/4$ interferometer action—i.e. $|m| = 1/d$, $3/d$, $5/d$, etc. ($\cos\pi\frac{1}{2}dm = 0$).

Equation (6.29) should strictly contain an additional factor $\sin\pi\frac{1}{2}dm/\pi\frac{1}{2}dm$ due to the width $W_1 = \frac{1}{2}d$ of the east–west antenna, which has been ignored so far. This will make little difference to the shape of A_c but will help to cut down the far responses in A_s. If the antennas are made up from identical cylindrical paraboloids or some other type of aperture with tapered grading, then the width of the antennas must be larger than $\frac{1}{2}d$. A further increase is needed if multi-beam observations are to be made (pages 140–1).

A great advantage of the grating T antenna is that it avoids the long continuous north–south aperture in which the phase of each dipole must be controlled individually. Instead, the grating allows the electrically much simpler combination of mechanical and electrical steering to be used.

Another advantage is that the same mechanical construction can be used for the north–south units as for the east–west antenna, which makes it much easier to maintain accurate phase relations between the two component antennas.

6.2.3. *The ring antenna*

If a ring were used as a power response telescope, it would have high sidelobes with respect to the main response. The Fourier transform of the grading $g(\rho_1) = 1$ for a uniform ring was calculated in section 2.43

$$f(0, m) = 2\pi\rho d\rho_1 J_0(2\pi\rho_1 m), \tag{2.48}$$

where $d\rho$ is the width of the ring. Its effective area is given by

$$A(0, m) = A_{\max}[J_0(U)]^2 \tag{6.30}$$

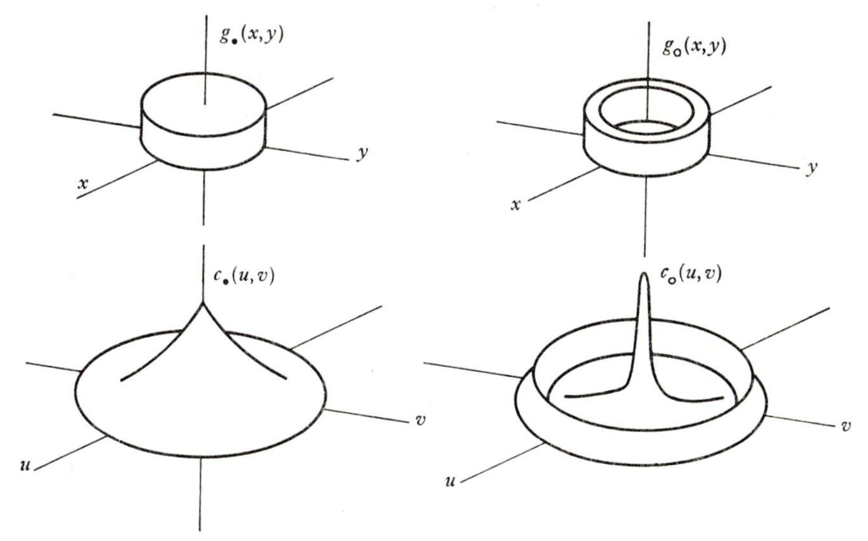

Figure 6.18. The gradings $g_\bullet(x, y)$ and $g_\circ(x, y)$ and the transfer functions $c_\bullet(u, v)$ and $c_\circ(u, v)$ of a uniform circular aperture and a ring aperture.

where $U = 2\pi\rho m$. This may be compared with the expression

$$A(0, m) = A_{\max}\left\{2\frac{J_1(U)}{U}\right\}^2 \tag{6.31}$$

for a uniformly graded filled circular aperture (see (2.53)).

The first sidelobe of a thin ring antenna has an amplitude of 16 per cent (power) of the main response, the second 9 per cent, and the third 6 per cent, etc. With such large sidelobes the antenna is obviously of very limited use. It has, however, the same range of spacings as a filled circular aperture of the same diameter. We see in Figure 6.18 that the weighting of the spectral components for the filled aperture and a ring antenna is quite different. In order to bring the transfer function $\mathbf{c}(u, v)$ for the ring antenna closer to that for the filled circular aperture it is necessary to reduce the size of the central peak and produce greater tapering of the higher components. With the cross and T antennas the transfer function has been made to approach that of a filled aperture (*a*) by dividing the antenna structure into two sections and correlating the outputs of the two sections and (*b*) by uneven weighting of different parts of each section of the antenna (i.e. tapering). With a ring array this cannot be done in such a simple way. One possibility is to modify the observations after they have been completed, as we discussed in section 6.1.3. The restoring function $R(l, m)$ that must be applied to the brightness-temperature distribution measured by a filled aperture is derived from (6.20)–(6.23).

$$R(l, m) \mathscr{F} r(u, v)$$

$$\mathscr{F} \left[\frac{\mathbf{c}_{\bullet}(u, v)}{\mathbf{c}_{\circ}(u, v)}\right], \tag{6.32}$$

where $c_{\bullet}(u, v)$ and $c_{\circ}(u, v)$ are the transfer functions of a filled aperture $A_{\bullet}(l, m)$ and a ring aperture $A_{\circ}(l, m)$ respectively, i.e.

$$\mathbf{c}_{\bullet}(u, v) \mathscr{F} A_{\bullet}(l, m)$$

$$\mathscr{F} \left[A_{\max}\left(2\frac{J_1(U)}{U}\right)^2\right], \tag{6.33}$$

$$\mathbf{c}_{\circ}(u, v) \mathscr{F} A_0(l, m)$$

$$\mathscr{F} [A_{\max}(J_0(U))^2]. \tag{6.34}$$

The process of correcting a map of brightness distribution $T_{a0}(l, m)$ by smoothing it with the restoring function $R(l, m)$ or, alternatively, the process of Fourier transforming $T_{a0}(l, m)$, multiplying this by the function $r(u, v)$ and then retransforming is tedious.

Another possibility that has been explored and has been applied at the Culgoora radio heliograph in Australia (Plate 3, facing p. 96) is the use of a process roughly analogous to the phase-switching employed with rectangular correlation systems such as the cross antenna.(611) With the cross antenna the relative phase of the two arms is repeatedly changed from an in-phase to an out-of-phase relation. The difference of the antenna responses in the two positions gives the required antenna pattern for the system. What has to be done with a ring antenna is to phase modulate the ring in a number of different ways and add or subtract the different responses with different weightings. This process has been called J^2 synthesis. It makes use of the fact that, if the current grading around a circular ring is such that the total phase change is $2\pi k$ radians then, corresponding to (6.30) for the uniform-phase grading, we have

$$A_k(0, m) = [J_k(U)]^2. \tag{6.35}$$

It is known that a circularly symmetrical response $A(0, m)$ can be expressed as a series

$$A(0, m) = \sum_{k=0}^{\infty} t_k A_k(0, m) \tag{6.36}$$

i.e. we have
$$A(0, m) = \sum_{k=0}^{\infty} t_k [J_k(U)]^2. \tag{6.37}$$

If, for example, we choose

$$\frac{A(0, m)}{A_{\max}} = 2\frac{J_1(U)}{U} \tag{6.38}$$

we can find the coefficients t_k to satisfy (6.37) and (6.38). Alternatively, other suitable forms for $A(0, m)$ may be chosen.

The only ring antenna that has been built to date takes the form of a discontinuous ring of steerable paraboloidal reflecting antennas. The discontinuity in the ring produces sidelobes analogous to the grating lobes of a grating cross and the antenna is suitable only for the observation of strong isolated sources.

The principal virtue of a ring compared with any of the rectilinear arrangements is similar to that of a filled circular aperture compared with a rectangular one; the sidelobe responses have no favoured directions. This means that the sidelobes are both smaller and more widely distributed in angle than with the cross or T antenna. The main disadvantages are that the process of forming a satisfactory antenna pattern is very complicated and that the antenna is essentially fixed in size; it cannot be easily extended in the way in which a cross or T or a rotational synthesis antenna can.

6.3. Various problems involved in forming a radiotelescope from a number of separate antennas

Unfilled-aperture antennas are usually very large, and radio energy is collected at a number of well-separated points. This energy has to be conveyed over considerable distances to a radio receiver. Unlike a steerable parabolic dish, most unfilled-aperture antennas are designed to receive energy from sources that are not, in general, normal to the plane of the aperture of the antenna. The energy received at different parts of the antenna, which is inclined to the wavefront, is required to be additive from some direction in space; to achieve this, the waves arriving at different parts of the antenna have to be delayed by unequal amounts so that they will reach the receiver simultaneously.

Generally, transmission lines are used to carry the energy from different parts of the antenna to a central point and, at the central point, variable lengths of transmission line may be used to introduce the necessary compensating time delays into the different contributions. This introduces problems that are not present in an antenna such as a parabolic dish, in which the energy is conveyed to the receiver almost entirely by an air path and in which the time delays are the same for all parts of the intercepted wavefront that is parallel to the aperture plane of the antenna.

Transmission lines are made from metallic conducting material or some dielectric material or, more usually, they include both conducting and dielectric materials. The materials introduce power losses and, in addition, have characteristics that vary with both temperature and frequency. These variations give rise to some of the main difficulties in the operation of a very large radiotelescope.

6.3.1. *Attenuation of power flowing along the transmission lines* is produced, as a rule, by resistive loss in the conducting material of the transmission line. Losses in modern dielectrics and radiation losses are usually smaller than those in the metal.

Since the useful conducting material in a transmission line is the part into which the current penetrates, and since the thickness of this layer of materials is inversely proportional to $\nu^{-\frac{1}{2}}$ the attenuation per unit length of line produced by metallic losses is proportional to $\nu^{-\frac{1}{2}}$ provided that the thickness of the metal is considerably greater than the penetration depth of the current. This attenuation, when it occurs before amplification of the signals, reduces the signal level with respect to the noise in the receiver and, in addition, introduces thermal noise (Chapter 8). In large antenna systems in

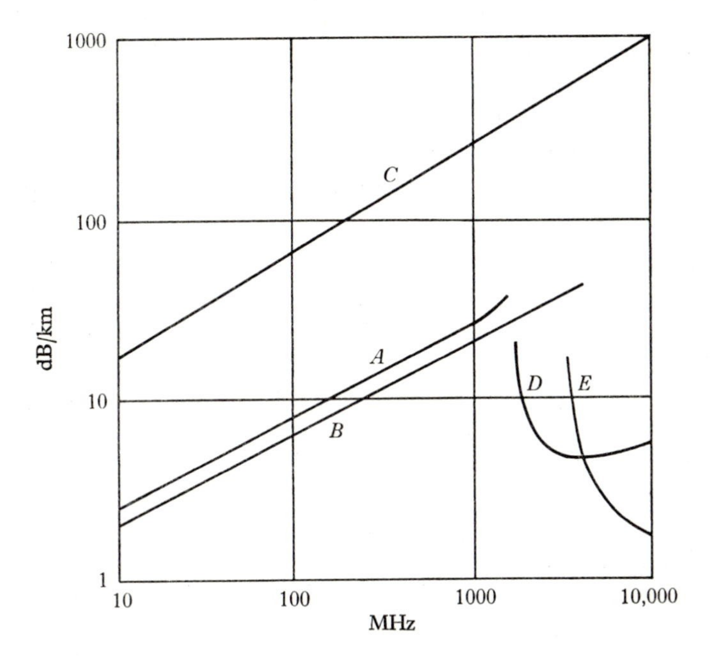

Figure 6.19. Approximate attenuation characterization (in dB/km) over a range of frequencies of: *A*, Balanced twin-wire line, conductors 4 mm diam. in air. *B*, Coaxial line, air dielectric. Outer conductor 4 cm diam. *C*, Typical solid dielectric coaxial line. Outer diam. 1 cm. *D*, 10 cm diam. circular waveguide, TE 11 mode. *E*, Same waveguide, TE 01 mode.

which the length of transmission line may be of the order of 1 km or more the attenuation of signals is very great (Figure 6.19), and it is essential to introduce preamplifiers at each main component part of the antenna. Common practice is to change the signal frequency to a lower frequency at these preamplifiers and to transmit the signals back to the receiver in an intermediate frequency band which may lie somewhere in the range from 1 to 60 MHz. In doing this, we may use a smaller and cheaper transmission line than would otherwise be necessary. On the other hand, the heterodyning signal, which has to be conveyed from a central location to each of the preamplifiers, is at a high frequency and suffers considerable attenuation. Hence, a high-power heterodyne oscillator is required. Alternatively, the heterodyne signal may be sent out at a lower frequency and the frequency be multiplied at each preamplifier. In any case, the attenuation of the heterodyne signal does not produce any deterioration of the signal with respect to receiver noise.

6.3.2. *Electrical steering*

This is required in a multi-element telescope, and involves a number of problems connected with receiver bandwidth and with the phase stability of the system.

We have shown in Chapter 2 that, with a single frequency signal, a linear phase shift along the antenna will shift the maximum of the antenna response to any required direction. If τ is the difference in the time of arrival of a wavefront at two different parts of the antenna, then at a single frequency ν_0 this difference represents a phase-difference $\phi = 2\pi\nu_0\tau$ which may be compensated by a phase shift ϕ_s' in one signal path, where

$$\phi_s = 2n\pi + \phi_s' \tag{6.39}$$

and n is an integer.

This simple method of steering is possible only when the product of the time difference and the bandwidth ($\tau\,.\,\Delta\nu$) is less than unity (section 5.5). To overcome the limitations of electrical steering to relatively narrow bandwidth signals we may introduce artificially a delay τ in the signal path from one antenna to the receiver, so that the voltages induced by the wavefront at the two antennas will arrive at the receiver simultaneously. This completely frequency-independent method of steering an array of antennas is not usually possible, because the accuracy required in adjusting the continuously variable delay line is too difficult to attain. What is more usual is to use a combination of the single-frequency and the frequency-independent methods, which is adequate for a band of frequencies. With this technique, time delays are added in discrete lengths, and a variable phase change is employed to keep the centre-frequency components of the received signals in the correct phase. For convenience the time-delay steps are added after the signal has been converted to a lower frequency by means of a heterodyne voltage. We shall consider this process quantitatively.

Let H and I be the *angular* frequencies of the heterodyne oscillator and the signal after conversion. Then the corresponding original *angular* signal frequencies are $(H+I)$ and $(H-I)$. Let us assume first that by means of a filter at each antenna only the component of angular frequency $(H+I)$ is admitted.

We use the symbols ${}_sV_1$ and ${}_sV_2$ for the signal voltages (from the antennas) that appear at the frequency changers of the receivers, ${}_iV_1$ and ${}_iV_2$ for the corresponding signals after conversion and after the delay has been added. ϕ_H is the phase difference between the voltages of the heterodyne oscillator at the two receivers, τ_s (sec) is the signal delay of the wavefront (Figure 6.20) and τ_i is the added delay in one receiver.

$$\left.\begin{aligned} {}_sV_1 &= {}_sa_1 \cos [(H+I)t+(H+I)\tau_s/2], \\ {}_sV_2 &= {}_sa_2 \cos [(H+I)t-(H+I)\tau_s/2] \end{aligned}\right\} \tag{6.40}$$

and, after multiplication with the heterodyne oscillator, and passing through filters and delays

$$\left.\begin{aligned} {}_iV_1 &= {}_ia_1 \cos [It+H\tau_s/2-\phi/2+I(\tau_s-\tau_i)/2], \\ {}_iV_2 &= {}_ia_2 \cos [It-H\tau_s/2+\phi/2-I(\tau_s-\tau_i)/2]. \end{aligned}\right\} \tag{6.41}$$

If the two voltages are multiplied together, then the product of ${}_iV_1 \cdot {}_iV_2$ taken over a cycle is

$$\langle {}_iV_1 \cdot {}_iV_2 \rangle = \text{const.} \cos \{H.\tau_s-\phi+I(\tau_s-\tau_i)\} \tag{6.42}$$

which is maximum when

$$H.\tau_s-\phi+I(\tau_s-\tau_i) = 0. \tag{6.43}$$

For any particular direction of arrival of a wave, i.e. for any value of τ_s and for one value of I, (6.42) can be made maximum by adjusting either ϕ or τ_i. However, we can satisfy (6.42) for any value of I (i.e. an infinitely wide band of frequencies) by making $\tau_s-\tau_i = 0$ and $\phi = H\tau_s$. Hence, to maximize the signal two adjustments are required, that of τ_i and ϕ_i for each value of τ_s. Had we compensated for τ_s *before* frequency change so as to make the effective value of τ_s equal to 0, then by fixing $\phi = 0$ and $\tau_i = 0$ we should have required only one adjustment, the compensating of the radio frequency delay, in order to maximize the signal.

Equation (6.42) does not give the overall receiver response in terms of a band of frequency. To get this we must introduce the normalized spectral sensitivity $b'(\nu')$ of the receiver (section 5.5) and its Fourier transform $B(\tau)$. τ is the time delay between the

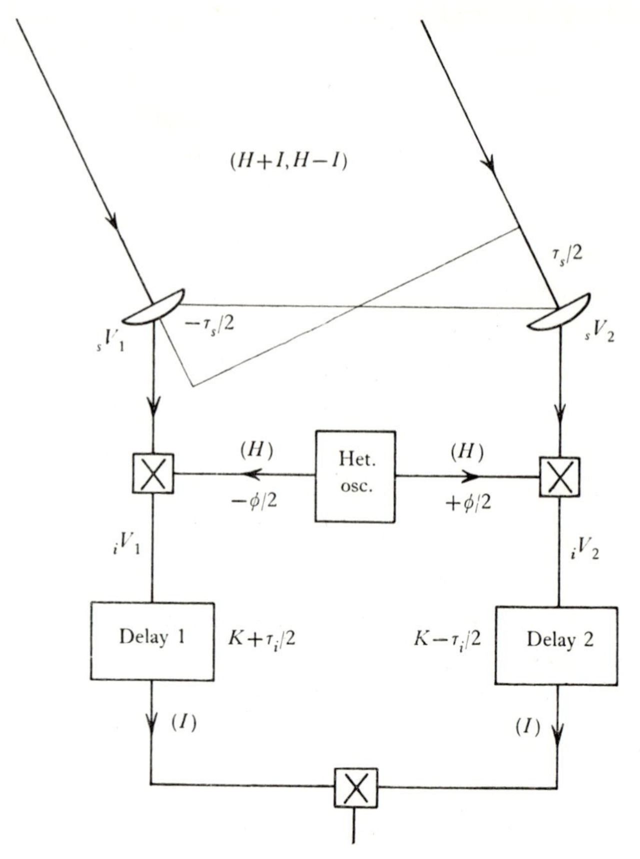

Figure 6.20. Compensation by delay τ_i in the intermediate frequency channel for a difference τ_s in the time of arrival of signals at two spaced antennas.

signals arriving by the two paths to the correlator and is equal, therefore, to $(\tau_s - \tau_i)$. The expression $B(\tau_s - \tau_i)$ provides the envelope function (fringe-washing function) to the expression in (6.42). For a band of frequencies, therefore, we have:

$$\text{Receiver response} = a \,.\, B(\tau_s - \tau_i) \,.\, \cos\{H\tau_s - \phi + I(\tau_s - \tau_i)\}$$

$$= a \,.\, B(\tau_s - \tau_i) \,.\, \cos\{(H+I)\tau_s - \phi - I\tau_i\}. \quad (6.44)$$

As an example, if we have a rectangular pass band $\Delta I = 2\pi\Delta\nu'$, then $B(\tau_s - \tau_i)$ becomes

$$B(\tau_s - \tau_i) = 2\sin\{\Delta I(\tau_s - \tau_i)/2\}/\Delta I\,.\,(\tau_s - \tau_i) \qquad (6.45)$$

(Figure 6.21 *a* and *b*).

In this discussion we have stated that the signal frequency $(H+I)/2\pi$ was accepted and the image frequency $(H-I)/2\pi$ was

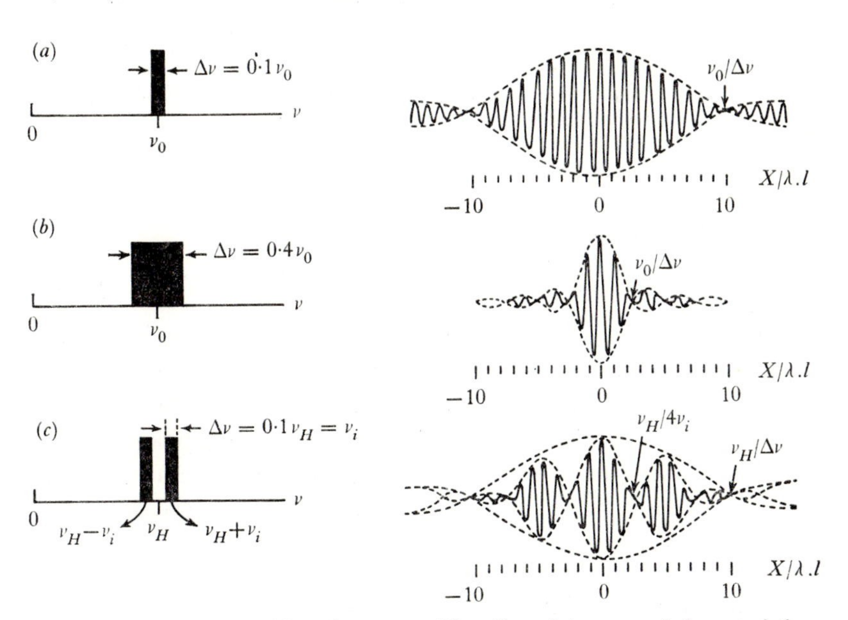

Figure 6.21. Broad-band interferometry: The effect of the spectral characteristic of the received signal on the telescope response (effective area). (*a*) A rectangular frequency band with a width equal to 0·1 of the centre frequency. (*b*) 0·4 of the centre frequency, and (*c*) Signal and image both accepted and each 0·1 of centre frequency in width. *X* is the length (metres) of the interferometer baseline.

filtered out. We shall now consider what happens when the image frequency is accepted.

For the angular frequency $(H-I)$ we have instead of (6.41) the following:

$$\left.\begin{aligned} {}_iV_1' &= {}_ia_1\cos[It - H\tau_s/2 + \phi/2 + I(\tau_s - \tau_i)/2],\\ {}_iV_2' &= {}_ia_2\cos[It + H\tau_s/2 - \phi/2 - I(\tau_s - \tau_i)/2] \end{aligned}\right\} \qquad (6.46)$$

or

$$\langle {}_iV_1' \cdot {}_iV_2' \rangle = \text{const.}\cos\{H\tau_s - \phi - I(\tau_s - \tau_i)\} \qquad (6.47)$$

and the conditions for maximizing (6.47) are the same as for those obtained in (6.43), i.e. $\phi = H\tau_s$ and $\tau_s = \tau_i$. The effect of the bandwidth of the receiver is practically the same as before, but it must be remembered that $b'(\nu')$ is measured in the reverse directions since the pass band of the image band is inverted with respect to that of the signal band. If, for simplicity, we assume that the pass band is symmetrical, then we have the same $B(\tau_s - \tau_i)$ as before and

$$\begin{aligned}\text{Receiver response} &= a \,.\, B(\tau_s - \tau_i) \,.\, \cos\{H\tau_s - \phi - I(\tau_s - \tau_i)\}\\ &= a \,.\, B(\tau_s - \tau_i) \,.\, \cos\{(H-I)\tau_s - \phi + I\tau_i\}.\end{aligned} \tag{6.48}$$

If we add the power responses of (6.44) and (6.48) we have

Power response of signal and image

$$= 2aB(\tau_s - \tau_i)\cos(H\tau_s - \phi)\cos[I(\tau_s - \tau_i)]. \tag{6.49}$$

This response, unlike that of the single-band responses, is doubly periodic, one frequency being proportional to H and the other to I, as Figure 6.21*c* illustrates. It has one particularly interesting feature when compared with that of the single band: the phase of the rapidly varying component is no longer dependent on τ_i. This, therefore, need not be known with any accuracy and need be varied only to keep the response near maximum.

6.3.3. *Image formation*

A directional antenna may be regarded as an energy rejecting device as well as an energy collecting one. If it has only a single directional response it rejects the information coming from all but one region of the sky. If the antenna is divided into many small areas, then the energy accepted by each small area comes from a large region of the sky. Instead of simply adding the voltages from all of the elementary areas, which would result in a large part of the received energy being rejected (re-radiated), we may amplify the contributions separately and afterwards combine the voltages in different ways to form multiple responses, i.e. to form an 'image' rather than a simple response. It will be shown in Chapter 8 that, if the antenna is divided so that N separate and independent responses are obtained, then the effective, or 'scanning', sensitivity of the telescope is increased by a factor of $\sqrt{N}$. Unfilled-aperture antennas made up of a number of parts, each with its own amplifier, are particularly suited to image formation.

The oldest method of image formation is by means of multiple beams, first used at Nançay in France.(612) The radio energy from each of the component antennas of a grating was amplified, divided, and passed into a number of different phase-shifting networks. The output of one network was connected to the output of networks connected to each of the other component antennas to form a single response. The phase-shifting networks were arranged so that each single response was directed to a different part of the sky (within the field of view of the component antennas) and by this means many different pieces of directional information could be obtained simultaneously.

The maximum useful number of separate responses that can be obtained is equal to the number of independent spatial components which can be extracted from pairs of component antennas that make up the system. This will not be proved, but it can be illustrated by a line antenna which has uniform grading and is divided into N parts. The angular response of each of the component parts is N times wider than that of the array, and it would seem reasonable to suppose that, by introducing linear phase shifts, N independent beam positions can be found inside the envelope of the response of each element. We refer the reader to Chapter 4, which discusses some of the problems associated with the electrical steering of an antenna response; the main difficulty is that the sidelobe level for responses that are steered away from the optimum direction of the component antennas is higher than for those near the optimum direction.

The method has been extended to cross and T antennas.(608) If a cross is divided into $\sqrt{N}+\sqrt{N}$ elements or a T into $\sqrt{N}+\sqrt{N}/2$ elements, there are $N/2$ independent pieces of information available from the antenna, i.e. $N/2$ separate responses can be extracted. With a T antenna or other asymmetrical antennas there are considerable difficulties if all elements do not lie in the same plane. If, say, one arm of a T lies in a different plane from the other arm, then for each separate response of the system a different and accurately calculated phase connection must be made before the signals are correlated, otherwise a large unwanted sine response may be added to the required cosine response of the system.

Another method of image formation which has a number of advantages uses the technique employed in aperture synthesis in which the interference patterns, both sine and cosine, from pairs of component antennas are amplified and extracted separately.[605] The amplitudes of these components are recorded for all the spacings between pairs of antennas in the array. They represent the Fourier components of the brightness of that region of the sky that is 'seen' by each of the component antennas. After transformation the radio brightness distribution over the region is obtained.

The two methods of forming an image by multiple beams or by the method of Fourier synthesis do not differ in the amount of information extracted. The difference lies in the equipment required to process the information: the multiple beam method uses what could be called 'analogue computer' methods which involve groups of phase shifters and adders to turn the information into the brightness-angle form. With the second method this process takes place in (probably) a digital computer which may or may not form part of the radio telescope.

CHAPTER 7

APERTURE SYNTHESIS

7.1. The correlation interferometer

An interferometer measures one Fourier component of the brightness distribution over some region of the sky.[701] The radiotelescopes which we have discussed in previous chapters contain a range of (element) interferometer spacings. They measure simultaneously a particular combination of all the Fourier components within the capability of each instrument, and this combination contains the information about the radiation from one small region of the sky. The telescope rejects information about the radiation from other parts of the sky.

The essential feature of an aperture-synthesis telescope is that the Fourier components are measured not simultaneously but serially. The telescope consists of one variable spacing interferometer or an assembly of them. Measurements of the same area of sky are made sequentially with different spacings of the interferometer antennas. This implies (*a*) that only a small part of the telescope aperture is present at any time, which leads to a considerable economy in material, (*b*) that one part of the interferometer must be moved in space many times with respect to the other so that all necessary Fourier components of the source distribution may be measured, and (*c*) that the source of radiation and the medium between it and the telescope must not change during the complete course of the observations.

The basis of the method is found in (5.27), which gives the complex antenna temperature $\mathbf{T}_a$ of a correlation telescope in terms of the brightness temperature T_B and the complex effective area $\mathbf{A}(l, m)$. We apply this to an interferometer in the following way:

$$\mathbf{T}_a = \lambda^{-2} \int_{4\pi} T_B(l, m)\, \mathbf{A}(l, m)\, d\Omega. \tag{5.27}$$

The correlation telescope effective area is (5.19)

$$\mathbf{A}(l, m) = |\mathbf{A}(l, m)| \exp j\psi, \tag{7.1}$$

where $$|\mathbf{A}(l,m)| = 2[A_1(l,m)\,A_2(l,m)]^{\frac{1}{2}} \tag{7.2}$$

is the envelope pattern (5.21). For an interferometer where the electrical centres of the component antennas are separated by u wavelengths in the x-direction and by v wavelengths in the y-direction we have the path difference ψ radians from the source to the two antennas

$$\psi = 2\pi\nu_0\tau$$
$$= 2\pi(ul+vm) \quad \text{(radians)} \tag{5.8}$$

If we put the effective area of the interferometer into (5.27) and replace $d\Omega$ by $dl\,dm/(1-l^2-m^2)^{\frac{1}{2}}$ (eqn 2.5) we obtain the correlation temperature $\mathbf{T}_a(u,v)$ at the particular interferometer spacing (u,v).

$$\mathbf{T}_a(u,v) = \iint_{-\infty}^{+\infty} \frac{T_B(l,m)\,.\,|\mathbf{A}(l,m)|}{\lambda^2\,.\,(1-l^2-m^2)^{\frac{1}{2}}}\,.\exp\{j2\pi(ul+vm)\}\,dl\,dm. \tag{7.3}$$

This is a Fourier integral equation; hence we have a Fourier pair of functions:

$$\mathbf{T}_a(u,v) \,\mathscr{F}\, \frac{T_B(l,m)\,|\mathbf{A}(l,m)|}{\lambda^2(1-l^2-m^2)^{\frac{1}{2}}}. \tag{7.4}$$

The formal integration limits $\pm\infty$ do not change the value of the integral if, as before, we define $T_B(l,m) \equiv 0$ 'outside' the horizon, i.e. for $l^2+m^2 > 1$.

The measured correlation temperature $\mathbf{T}_a$ regarded as a function of the interferometer spacing is the Fourier transform of the expression in front of the exponential factor in the integral. A fixed spacing interferometer, therefore, measures one Fourier component of this 'weighted' brightness distribution. Further, we can obtain the required brightness distribution $T_B(l,m)$ from the inverse transform of the correlation temperature $\mathbf{T}_a$, which function can be measured with a variable spacing interferometer:

$$\frac{T_B(l,m)\,.\,|\mathbf{A}(l,m)|}{\lambda^2\,.\,(1-l^2-m^2)^{\frac{1}{2}}} = \iint_{-\infty}^{+\infty} \mathbf{T}_a(u,v)\,.\exp\{-j2\pi(ul+vm)\}\,du\,dv. \tag{7.5}$$

This is the fundamental equation of aperture synthesis: it shows that the sky brightness distribution within the envelope pattern (i.e. for $|\mathbf{A}| \neq 0$) can be calculated from the measurements $\mathbf{T}_a(u,v)$ with a variable spacing interferometer.

The brightness temperature distribution $T_B(l, m)$ and the envelope pattern $|\mathbf{A}(l, m)|$ are both real functions and it follows (Appendix 2) that its Fourier transform must be an even (sometimes called Hermitian) function, i.e. that

$$\mathbf{T}_a(-u, -v) = \mathbf{T}_a^*(u, v). \tag{7.6}$$

This means that, if $\mathbf{T}_a(u, v)$ is known, there is no need to measure $\mathbf{T}_a(-u, -v)$ or, in other words, only two of the quadrants in the u, v plane need be explored. This is obvious when the two antennas are identical: the same information is available if they exchange places. The extra quarter wave delay in the sine receiver, however, will, in effect, have been shifted to the other side, which changes the sign of the sine output (imaginary part of $\mathbf{T}_a$). That is the reason why the complex conjugate appears in (7.6).

7.2. Methods of synthesis

7.2.1. *The movable pair of antennas*

In early partial-synthesis measurements by interferometers at Cambridge one antenna was moved with respect to the other to produce different spacings (u, v) along a straight line.[702] A line antenna could thereby be synthesized. Because of technical difficulties the relative phase of the signals arriving at the two antennas could not be measured and only the resultant amplitudes were determined. This incomplete information could lead to a determination of $T_B(l, m)$ only if a simplifying assumption were made about the measurements; the simplest assumption was that only the cosine component was being measured, i.e. the source was symmetrical.

An important step in approaching true synthesis was the introduction of accurately placed steel tracks along which an antenna could be moved. An arrangement with two tracks placed at right angles was introduced in France and in California (Figure 1.10) so that a two-dimensional aperture could be synthesized.[703–4] A great number of separate measurements is required with this system to synthesize a large aperture and at each spacing phase calibration of the antenna is required. The technical difficulties are so great that it was many years before a complete and true synthesis of a region was attempted. The first true syntheses were effected by means of antennas that used one or both (Figure 7.1, Plates 9 and 14, facing p. 96) of the component antennas in the form of a long fixed array so that a two-dimensional map was achieved by means of a one-dimensional synthesis procedure.(705–7) The antenna shown in Figure 7.1 is of a meridian transit type and consists of a long east–west line array and a movable smaller north–south antenna (Plate 14, facing p. 96). The number of separate measure-

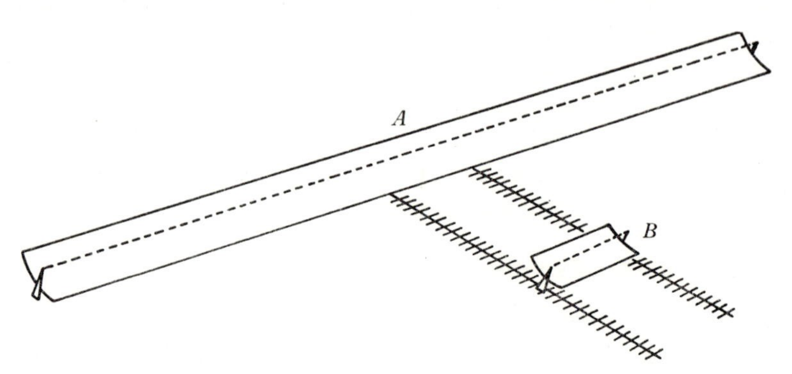

Figure 7.1. A synthesis telescope which has one movable antenna and one long fixed antenna. This is one of the first complete synthesis instruments (both phase and amplitude measured) built at Cambridge. An equivalent telescope is made if half of A is removed and the track of B is extended for an equal distance on the opposite side of A.

ments required to synthesize the aperture is reduced drastically compared with the number required when both antennas are movable and small. The other system shown in Plate 9 was the prototype of antennas which used the Earth's rotation to synthesize an aperture.(708) We shall describe such antennas in the next section.

Grading and effective area of the synthesized telescope. It is clearly impossible to make measurements at all spacings out to the integration limits of $\pm\infty$. The effect of this is the same as that of a limited aperture of a non-synthesis antenna: the angular resolution is limited.

Let $g_s(u, v)$ be the weighting or *grading of the synthesized aperture*, defined to be zero at all spacings (u, v) for which no measurements have been made. The product

$$\mathbf{T}_a(u, v) \cdot g_s(u, v) \tag{7.7}$$

is then, in contrast to $\mathbf{T}_a$ itself, known at all values of (u, v) and its inverse Fourier transform can be calculated. Replacing $\mathbf{T}_a$ in (7.5) by the product $\mathbf{T}_a . g_s$ will, of course, change the result, i.e. the weighted sky brightness temperature distribution on the left-hand side. The convolution theorem in Fourier analysis shows that, if (7.5) is valid, then

$$\frac{T_B(l,m) . |\mathbf{A}(l,m)|}{\lambda^2 . (1-l^2-m^2)^{\frac{1}{2}}} \star \Gamma(l,m)$$

$$= \iint_{-\infty}^{+\infty} \mathbf{T}_a(u . v) . g_s(u,v) . \exp\{-j2\pi(ul+vm)\}\, du\, dv, \qquad (7.8)$$

where $\Gamma(l,m)$ is the Fourier transform of the synthesis grading $g_s(u,v)$ and the symbol $\star$ indicates the smoothing integral of the two expressions on the left-hand side.

The direct result of the synthesis measurements and calculations is, therefore, a map or 'image' of the radio brightness distribution within the envelope pattern weighted by

$$\lambda^{-2}(1-l^2-m^2)^{-\frac{1}{2}}|\mathbf{A}(l,m)|;$$

this image has been smoothed so that the angular resolution corresponds to observations with a 'synthesized beam' $\Gamma(l,m)$. This beam is the Fourier transform of the synthesis grading $g_s(u,v)$ and therefore has the same shape as the power pattern of any telescope whose transfer function is proportional to $g_s(u,v)$; it also has the same shape as the *field* pattern of a single antenna whose grading is $g_s(x,y)$, that is, one whose dimensions are twice as large as the largest *spacing* used in the measurements. Usually g_s is chosen so as to taper smoothly towards the edges of the measured (u,v) region. This has the effect of reducing the otherwise rather large side-lobes. The normalized transform of the synthesis grading

$$\{\Gamma(l,m)\}_{1\cdot 0}$$

is the power pattern of the 'synthesized' telescope. At the centre of the image the effective area of the synthesis interferometer is

$$|\mathbf{A}|_{\max} = 2(A_{1\max} A_{2\max})^{\frac{1}{2}} \qquad (7.9)$$

and we can define an *effective area of the synthesized telescope* by

$$A_{\text{synth}}(l,m) = 2(A_{1\max} A_{2\max})^{\frac{1}{2}} . \{\Gamma(l,m)\}_{1\cdot 0}. \qquad (7.10)$$

The map or image produced by the telescope as given by the left-hand side of (7.8) can now be written

$$\text{L.H.S.} = \text{const.}\ \lambda^{-2}\left[T_B(l,m)\,.\left\{\frac{|\mathbf{A}(l,m)|}{(1-l^2-m^2)^{\frac{1}{2}}}\right\}_{1\cdot 0}\right]\star A_{\text{synth}}(l,m) \tag{7.11}$$

which, apart from the extra weighting factor in the curved brackets, has the same form as the expression for the map obtained by measurements with a conventional telescope whose beam is electrically steered (5.52). It is directly comparable with the expression we would get for observations with a grating or a compound grating antenna where the main beam in an analogous way may be steered electrically so as to explore a region of sky which is weighted by an envelope pattern (determined by the beams of the individual units in the arrays). The equations for synthesis are, in fact, qualitatively identical with the corresponding equations for a compound grating antenna, because measurements are made at a finite number of discrete spacings (u, v). The equivalent synthesized compound grating consists of the fixed antenna and an array in which there is one antenna, identical with the moving antenna at each point (u, v) which has been occupied during the synthesis measurements. It shares with ordinary compound interferometers the inherent difficulty of measuring the zero spacing component with two antennas which must not overlap (section 6.1.2), but it has this advantage, that the individual antennas in the synthesized array may be as close as we like, since neighbouring units will not be present at the same time.

If the synthesis spacings (u, v) are chosen in the simplest possible way, i.e. as a rectangular grid of points with regular intervals δu and δv in the u and v directions respectively, then the Fourier integral on the right-hand side of (7.8) is replaced by the corresponding series. For simplicity, write $T_{\text{map}}(l, m)$ for the whole expression on the left-hand side of (7.8). Then

$$T_{\text{map}}(l,m) = \sum_{v=-V}^{v=+V}\ \sum_{u=-U}^{u=+U} \mathbf{T}_a(u,v)\,.\,g_s(u,v)\,.\,\exp\{-j2\pi(ul+vm)\}\,\delta u\,\delta v, \tag{7.12}$$

where $$T_{\text{map}}(l,m) = \left[\frac{T_B(l,m)\,|\mathbf{A}(l,m)|}{\lambda^2(1-l^2-m^2)^{\frac{1}{2}}}\right]\star\Gamma(l,m)$$

and where the limits V and U must be large enough to contain the whole measured region in the u, v plane. Outside the measured region we have, by definition, $g_s(u,v) = 0$.

The double sum is best computed in two steps. Reversing the order of summation, the computer forms a function $Q(u,m)$ for all measured u and at intervals of m as desired on the synthesis map:

$$Q(u,m) = \sum_{v=-V}^{v=+V} \mathbf{T}_a(u,v) \,.\, g_s(u,v) \,.\, \exp\{-j2\pi vm\} \,.\, \delta v. \quad (7.13)$$

This is equivalent to forming the one-dimensional Fourier series development of each column (constant u) separately. In the second step the rows (constant m) are developed separately to give the desired map. If the two measured quadrants in the (u,v) plane are those with positive values of u we can write

$$\begin{aligned} T_{\text{map}}(l,m) &= \sum_{u=-U}^{u=+U} Q(u,m) \,.\, \exp\{-j2\pi ul\} \,.\, \delta u \\ &= -Q(0,m) + 2\sum_{u=0}^{u=+U} \text{Re}\,[Q(u,m) \,.\, \exp\{-j2\pi ul\}] \,.\, \delta u. \end{aligned} \quad (7.14)$$

The last step follows from $Q(-u,m) = Q^*(u,m)$, which can be derived from (7.6) and (7.13). It should be remembered in all these summations that $\mathbf{T}_a = T_c + jT_s$ and Q, as well as the exponential factors, will, in general, be complex functions. Computer programmes for Fourier series calculations can be made very fast and efficient.[709]

Grating response and the field of view. Equation 7.12 shows the form taken by (7.8) when observations are made at finite spacing intervals δu, δv. The main difference between (7.8) and (7.12) is that the series in (7.12) is periodic. $T_{\text{map}}(l,m)$ repeats itself each time the products ul or vm change by unity—i.e.

$$\delta l = 1/\delta u \quad \text{or} \quad \delta m = 1/\delta v. \quad (7.15)$$

The left-hand side of (7.12) contains an envelope function $|\mathbf{A}(l,m)|$ of the component antennas of the telescope. If the repetition intervals δl and δm are made large enough to ensure that only one response falls inside the envelope $|\mathbf{A}(l,m)|$, the observations are unambiguous. Hence, we may specify that the minimum value of δl and δm should be equal to the total angular width of $|\mathbf{A}(l,m)|$

measured from zero to zero. If, for example, the component antennas are uniformly graded square apertures each side of which is L wavelength long, then we find from (2.45) that δl and δm must not be smaller than $2/L$—that is, δu and δv must not be larger than $L/2$. If the elementary antennas, instead of being square and uniformly graded apertures, are circular and non-uniformly graded, δu and δv must be reduced to about $L/3$. This corresponds to the relation between the required antenna diameter and the element spacing in the compound grating antenna discussed in section 6.33. If the image area is restricted, or if sidelobe responses from sources outside the image region are not troublesome, then the spacing between measurements may be increased.

7.2.2. *The use of the rotation of the Earth*

The principle of rotational synthesis. The Earth's rotation has been used from the earliest days of radio astronomy for observations of a strip of the sky by means of a fixed radio antenna.[(710)] It was used also in the earliest interferometer observations which made use of sea reflection to provide a second path between a source and the antenna.[(711)]

When arrays of fully steerable paraboloidal antennas were used to follow a single source during the time when it was above the horizon, another property of the Earth's rotation was utilized. This useful property of the Earth's rotation is seen in projection from any fixed direction in space: most straight lines on the Earth appear to change both in length and in *direction* during the *rotation* of the Earth (Figure 1.11).

The exceptions are lines which lie parallel to the Earth's axis. These change neither in apparent length nor in direction. Another exception is a line parallel to the Earth's equator when viewed from a source in the equatorial plane of the Earth, i.e. at zero declination. This changes only in apparent length. If a line antenna lying in an east–west direction on the Earth is viewed from the celestial pole which lies in the same hemisphere, it will appear to rotate uniformly, and during twenty-four hours its orientation will change continuously over 360° (Figure 7.2*a*). Hence, we may synthesize an aperture in which all orientations are present and all distances up to the length of the antenna are found. In fact, as we have shown in section 7.1, all *necessary* orientations are found in twelve hours during which the antenna rotates through 180°.

The same synthesis may be effected by a pair of antennas for which the spacing may be changed step by step after each complete rotation of the Earth (Figure 7.2*c*).

Synthesis of an aperture by rotation of the Earth has many practical advantages over moving-antenna synthesis, but it has also

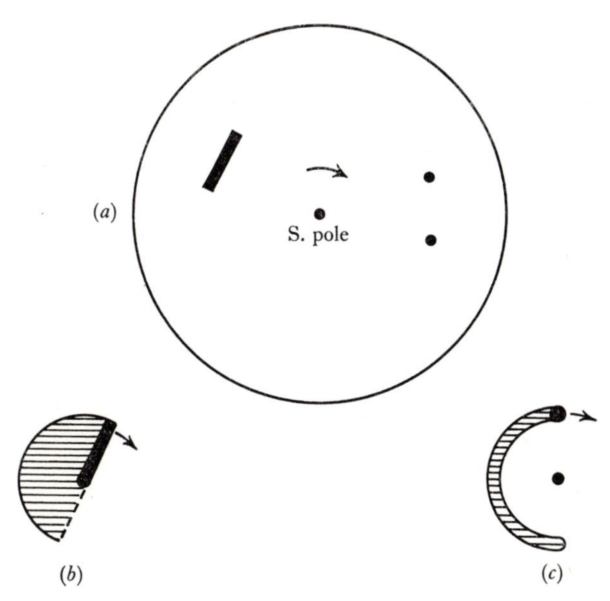

Figure 7.2. (*a*) View of an east–west line antenna and an east–west two-aerial interferometer seen from a point in space in the direction of the south celestial pole. (*b*) The range of apparent directions and spacings produced by a line antenna during twelve hours' rotation of the Earth. (*c*) The range of apparent directions and spacings produced by a two-aerial interferometer during twelve hours.

certain limitations. The synthesized pattern, though constant in shape when expressed in the (l, m) coordinates, becomes very elongated in angular measure for directions that lie near the celestial equator ($\delta = 0$); that is, the synthesized aperture, seen in projection from the direction of the source, is much smaller for low declinations. Figure 7.3 shows the range of apparent lengths and orientations of a line antenna lying horizontally on the Earth's surface and inclined to the Earth's axis at an angle χ when viewed from a source at declination δ during a period of twenty-four hours. As we have already stated, half of these directions are necessary: i.e. the line must be visible from the source for at least twelve hours. If the source lies *in the same hemisphere* as the antenna, then it is above the horizon for at least twelve hours. For the other hemisphere a complete range of scanning angles is not available.

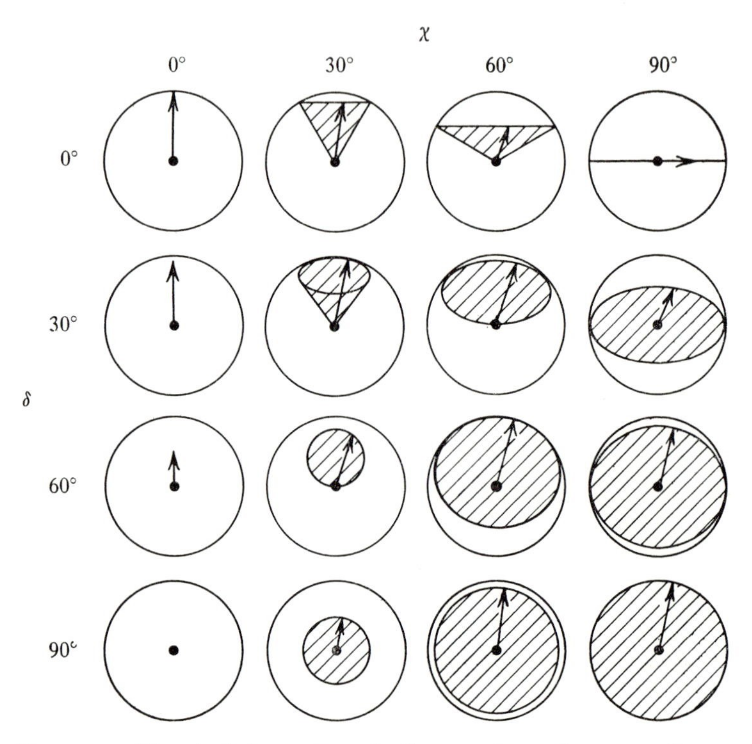

Figure 7.3. The shaded area in each diagram shows the range of apparent spacings and directions of a line antenna inclined at an angle χ to the Earth's axis and viewed during twenty-four hours from a point in space at an angle δ. For a horizontal east–west antenna $\chi = 90°$. For a north–south antenna χ is equal to the latitude ϕ at which the telescope is operating.

In Figure 7.3 χ (the angle between the axes of the antenna and of the Earth) is related to the latitude ϕ of the antenna on the Earth and its direction A with respect to the north–south direction by

$$\cos\chi = \cos A \cos\phi. \tag{7.16}$$

The apparent angle β that the line of the antenna makes with the

Earth's axis when viewed from the source in space is given by

$$-\cot\beta = \cos\delta\tan\chi\,\mathrm{cosec}\,H' + \sin\delta\cot H' \qquad (7.17)$$

where H' *is the sidereal time referred to the time when* $\beta = 0$. For an east–west line antenna $H' = (H-90°)$ and $\chi = 0$, where H *is the hour angle of the source* and (7.17) becomes

$$\cot\beta = \sin\delta\tan H. \qquad (7.18)$$

For a north–south line antenna we have $\chi = (90° - \phi)$, $H = H'$ and

$$-\cot\beta = \cos\delta\cot\phi\,\mathrm{cosec}\,H + \sin\delta\cot H. \qquad (7.19)$$

The ranges of apparent angle β shown in Figure 7.3 include parts in which the antenna lies in the Earth's shadow. The range of hour angles H during which the source is visible from the antenna lie between the hour angles $\pm H_H$ of rising and setting of the source. It can be shown that H_H is given by

$$\cos H_H = -\tan\delta\tan\phi. \qquad (7.20)$$

In Figure 7.3 we have considered the effect of the Earth's rotation on a line antenna. It is seen that, for observations of regions of the sky near the celestial equator, the rotational synthesis method when employed with a one-dimensional antenna is not satisfactory. It is obvious, however, that the method can be extended to two-dimensional arrays which need not have the restrictions imposed on a one-dimensional system. An example of a two-dimensional arrangement is shown in Figure 7.4(*a*). This consists of a north–south array which might, for example, be synthesized by a single antenna free to move along this line, with a single fixed antenna placed to the east or west of one end of this line. For simplicity the antenna is assumed to be at the equator and the source at zero declination. As the source travels from horizon to horizon, the projection of the single antenna moves from coincidence with the north–south array to a maximum separation at meridian transit of the source and then back again to zero. A continuous range of antenna spacings is obtained in this way, as 7.4*b* shows.

Many other two-dimensional arrangements are possible.

Types of rotational synthesis radiotelescope. Radiotelescopes which make use of the Earth's rotation in order to synthesize a two-dimensional aperture from an east–west line aperture must be capable of following a source for twelve hours. (This time may be reduced if two or more separate line-antennas in different orientations are employed.) During this twelve hours the antenna must be constantly receiving energy from the same region of the sky, i.e. from the same sources of emission. If the telescope consists, for example, of an array of antennas with each antenna mounted equatorially (having an axis of rotation parallel to the Earth's axis),

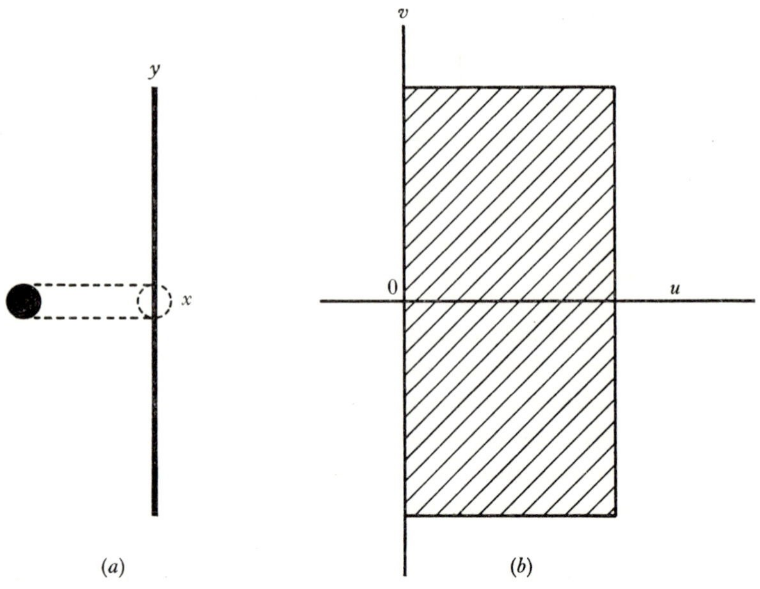

Figure 7.4. (*a*) A north–south line antenna at the equator of the Earth, with its output correlated with that of a single antenna placed east of the mid-point of the line antenna. (*b*) During six hours of observation all the spacings in the shaded area of the u, v plane are found.

the region from which energy is received by the individual antennas does not change, nor does the polarization of the received incoming radiation change during the twelve-hour course of the observations. If the antennas are not equatorially mounted, both the extent of the region and the polarization accepted may vary during the observations, and the conditions for synthesis of the region may not be met. All rotational synthesis instruments to date have consisted of fully steerable paraboloidal antennas equatorially mounted. The earliest of these, built in Sydney (Plate 9), consisted of two gratings at right angles.[708] Each grating consisted of a large number of equatorially mounted paraboloidal antennas capable of following a source (the Sun) for many hours during the day. A large range of angles with respect to the axes of the Sun were obtained during one day, and it was possible to use this to derive a two-dimensional brightness distribution over the Sun. Later systems with fixed-spacing antennas use compound grating interferometers rather than simple gratings and do not suffer from the inherent restrictions of the simple gratings.[605] The compound gratings are connected so that the various sine and cosine components of a Fourier series are extracted separately, as we have described in section 6.33, so that information from the whole field of the individual antennas is used at all times. With such a system a complete set of observations of the region is obtained in twelve hours (Plate 15, facing p. 96).

An alternative form of rotational synthesis consists of a simple interferometer in which one element is capable of being moved in only one direction (for example, east–west) with respect to the other. The Earth's rotation provides the necessary range of scanning angles, and the variable spacing between the two elements provides the necessary range of spacings. This double type of synthesis has sometimes been called 'supersynthesis'. The first antenna built specifically for this purpose, at Cambridge, in addition to having a fixed location antenna and a movable one on rails, uses a second fixed location antenna to provide a further range of spacings which follows on from those provided by the close pair of antennas.[712] In Holland this has been carried further by building a full grating of fixed location antennas with a pair of movable ones on a short rail-track (Plate 11, facing p. 96).[713] By this means a large number of separate observations can be made in twelve hours, and the number of days required to complete a set of observations for one region of the sky can be reduced considerably compared with the three-antenna arrangement.

We should point out that annular gaps in a synthesized antenna will produce secondary responses analogous to the multiple responses of a grating antenna. These secondary responses do not matter if we are studying small isolated sources. Hence, for such observations the spacing intervals δu between successive observations may be made much greater than the value given in (7.15).

The use of the Earth's rotation for synthesis of a two-dimensional aperture from a one-dimensional one has the great advantage that a large number of different bits of information are obtained without the necessity of relocating or readjusting or recalibrating the equipment between observations. Hence, the technique is much simpler than that needed for synthesis systems which do not make use of the rotation of the Earth. The rotational synthesis method, therefore, is more convenient than other methods for observation of sources in the same hemisphere as that in which the telescope is located and for sources that are not close to the celestial equator.

Special technical problems connected with rotational synthesis. The significant feature of rotational synthesis is that it involves many hours' tracking of a celestial source. During this time the system must respond in a constant way to different amplitudes and phase relationships of incoming signals. Since the ambient temperature is likely to change considerably during such a long period, the transmission lines and amplifying equipment must not be affected significantly by the changes in ambient temperature, or, if they are, then the changes must affect all signal paths equally.

Changes of the electronic equipment with ambient temperature changes can usually be reduced by suitable design, by means of temperature control of the critical parts of the equipment. They can, further, be made nearly equal by the use of similar equipment in each signal path so that differential effects, which are the only important ones, are minimized. The influence of transmission line changes can be reduced in the same way.

Changes that remain after these precautions have to be taken into account. They must be measured if they cannot be calculated. Calibration systems such as are required in all very large antenna arrays can be used to measure phase and amplitude changes in signal from each antenna to the correlation receiver. The best calibration is by use of a suitably located celestial radio source which is known to be small enough to be considered as a point source.[(714)] If the source has approximately the same declination and right ascension as the region being surveyed, the calibration takes into account all parts of the system, including errors in location of the individual antennas. Unfortunately, such suitable point sources are not always available for this purpose. Other calibration methods involve the introduction of signals from a uniform-phase (apart from 180° changes) calibration system which makes use of reflection or reversal of direction to produce a signal of constant phase.[715] Alternatively, a calibration signal may be introduced into each antenna from a central generator through a path of known, usually measured, electrical length. Such methods as these do not check the antenna positions and therefore are not as good as the point-source calibrations.

The second problem associated with rotational synthesis is that, during the course of the observations, the difference in the length of the signal paths between the source and any two component antennas varies continuously and by large amounts. Unless this is compensated for, the different frequencies within the pass band of the receiver will be affected differently by the changing position in the sky of the source. We have discussed in sections 5.32 and 6.32 the connection between signal delays and the usable receiver bandwidth. The power response of a receiver when connected to a pair of antennas at which signals arrive with a time difference τ_s, partly compensated by an added delay τ_i, was shown to be proportional to a function $B(\tau_s - \tau_i)$, the Fourier transform of the spectral sensitivity function $b(\nu')$ of the receiver. In a rotational synthesis we must keep $(\tau_s - \tau_i)$ sufficiently small so that $B(\tau_s - \tau_i)$ does not depart by more than a certain stipulated amount from its optimum value. If, for simplicity, we assume that the receiver has a rectangular pass band of width $\Delta I = 2\pi\Delta\nu'$, then (6.45) indicates that the receiver response is proportional to

$$B(\tau_s - \tau_i) = 2\sin\{\Delta I(\tau_s - \tau_i)/2\}/\Delta I\,.\,(\tau_s - \tau_i). \tag{6.45}$$

This function falls to zero when

$$|\tau_s - \tau_i| = 2\pi/\Delta I = 1/\Delta\nu'. \tag{7.21}$$

If we specify that the error must not exceed 5 per cent, the allowable difference $\tau_s - \tau_i$ is given by

$$|\tau_s - \tau_i| \ngtr 0{\cdot}175/\Delta\nu'. \tag{7.22}$$

For a telescope in which the signal and image are both received the allowable difference $(\tau_s - \tau_i)$ is smaller than this if the signal is to remain within the specified limits.

Equations (6.45) and (6.49) show that the normalized power response is given by

$$2\sin[\Delta I(\tau_s-\tau_i)/2]/\Delta I(\tau_s-\tau_i) \,.\, \cos[I(\tau_s-\tau_i)] \qquad (7.23)$$

and we specify that this must not fall below 0·95.

If we take the instance where I has the lowest possible mean value $= \Delta I/2$ so that the two bands adjoin, then (7.23) becomes

$$\sin[\Delta I(\tau_s-\tau_i)]/\Delta I(\tau_s-\tau_i) \not< 0{\cdot}95 \qquad (7.24)$$

or

$$|\tau_s-\tau_i| \not> 0{\cdot}087/\Delta\nu'. \qquad (7.25)$$

This tolerance in $\tau_s - \tau_i$ is the same as it would be if the bandwidth $\Delta\nu'$ had been doubled, which in fact it has been, because two bands each of width $\Delta\nu'$ are now being received simultaneously. The system has the advantage mentioned in Chapter 6 that the length of the delay elements added in the intermediate-frequency channel is not critical.

7.3. The use of aperture synthesis methods for extremely high resolution observations

For extremely high resolution observations the methods of aperture synthesis are invaluable. For such measurements it is usually not necessary, or even possible, to synthesize a complete aperture. Only enough observations are taken to provide the minimum necessary amount of information. Sometimes a determination of the size and position of small objects is sufficient.

Long base-line interferometry has been the main tool of radio astronomers, but valuable observations have been made by measuring the diffraction pattern on the Earth of the limb of the Moon when a celestial source is occulted by the Moon, or rises over the Moon's horizon. This lunar occultation method is somewhat analogous to the first interferometry measurements ever made by radio astronomers, in which the Earth's horizon rather than the Moon's was used to provide a diffraction pattern of a radio source.

7.3.1. *Long base-line interferometry*

With two antennas this requires that the signals, received at two points very far apart, be brought to a common point where they are correlated after suitable time delays have been added to one of the signals. When correlated a rapid interference pattern is produced if the source is not large enough to 'wash out' the fringes, i.e. if the brightness distribution has Fourier harmonics of sufficiently high order to be received by the widely spaced interferometer.

One method of combining the two signals is to bring one to the other by means of a radio link; for example, one received signal can be amplified and then used to modulate a radio-frequency carrier of much higher frequency, and the modulated wave can then be transmitted to the second antenna, where it is demodulated and correlated with the second signal. This method has been used at S. Michel in France. A more usual means, however, is to transmit a continuous wave signal to the two separated antennas, to use this signal to beat with the radio astronomical signals, and then to transmit the resultant beat-frequency (i.f.) signals to some common point. This method was used in Sydney in early experiments and has been developed into a very powerful tool by radio astronomers at Jodrell Bank and other places.[716–18]

For very great distances, up to the maximum obtainable on the Earth, the use of a common heterodyne signal has been replaced by two separate heterodyne oscillators of extremely high stability.[719–20] After frequency reduction the radio astronomical signals are recorded on magnetic 'video' tape and these tapes arc later correlated. Fringes at a spacing of $\frac{1}{1000}$ second of arc have been observed by this means.

There is another method, which can avoid the use of a common heterodyne signal. This method, called *post-detector correlation*, was developed at Jodrell Bank.[722] Signals are demodulated at each of the widely separated antennas and the demodulated signals are recorded and later correlated. In essence the system makes use of relative phase relationships within the pass band of each receiver, and these provide a correlated component when the two demodulated signals are combined. Post-detector correlation has been used to some extent in radio astronomical observations, but it is essentially a system which can be used only when the signal is strong compared with the noise level in the receivers.

In these long base-line observations it is usual to measure only the amplitude of the interference pattern so thet actual brightness distribution maps are normally not obtainable.

7.3.2. *Lunar occultations*

This provides an alternative method for obtaining high resolution observations.[723] Like the Lloyd's mirror type observations in the early days of radio astronomy it avoids the great technical difficulties of the two-antenna method, in that it requires only a single receiving antenna and a simple receiver. It is not possible, however, to increase the sensitivity of the system by time integration, since the occultation takes place very rapidly and is not repeated. Hence, a large steerable receiving antenna is needed, and the method was developed first around the large telescopes at Jodrell Bank and at Parkes.[724]

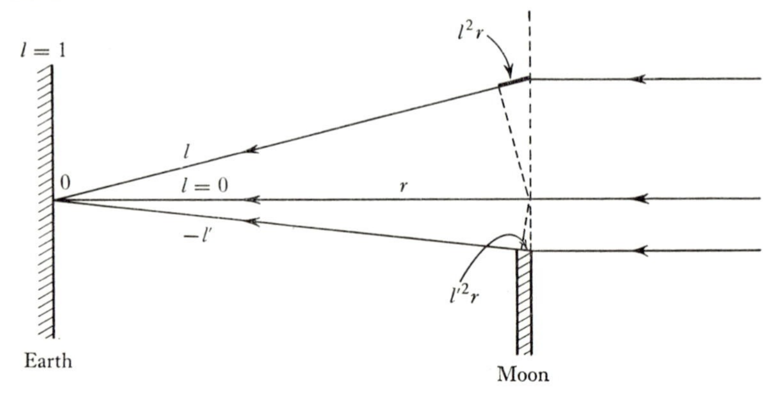

Figure 7.5. Diffraction at a straight edge (Moon) of rays from a distance source. The diffraction pattern is swept over the point O on the Earth by the movement of the Moon.

It can be explained briefly as follows: If we take a point source at infinite distance on the z-axis ($l = 0$) and replace the Moon by an absorbing screen the edge of which is parallel to the y-axis and at distance r from it and in a direction $l = -l'$ (Figure 7.5), then the resultant field $\mathbf{E}(l', 0)$, when the source is at a distance l' from the limit of the Moon, is given by

$$\mathbf{E}(l', 0) = \text{const.} \int_{-l'}^{1} \exp(-jql^2)\, dl, \tag{7.26}$$

where $q = (2\pi r/\lambda)$. The power flux corresponding to $\mathbf{E}$ is proportional to

$$\mathbf{EE}^* = \text{const.} \int_{-l'}^{1} \exp(jql^2)\, dl \int_{-l''}^{1} \exp(-jql^2)\, dl. \tag{7.27}$$

The differential of this power flux with respect to l' is

$$\frac{d}{dl'}(\mathbf{EE^*}) = f'(l', 0) \propto \exp(jql'^2)\int_{-l'}^{1} \exp(-jql^2)\, dl + \exp(-jql'^2)\int_{-l'}^{1} \exp(jql^2)\, dl. \quad (7.28)$$

The expression, when convolved with the one-dimensional (or strip) brightness $T_B(l', 0)$ of the source, gives the differentiated, observed occultation curve $T_a(l', 0)$. What is required is to find a restoring function $R(l, 0)$ which, when convolved with $T_a(l', 0)$, will give $T_B(l', 0)$. The process of restoration and its limitations will not be discussed here. The restoring function $R(l, 0)$ can be shown to be the expression given in (7.28) but as a function of $-l'$ instead of l',[725] i.e.

$$R(l, 0) = f'(-l', 0).$$

The process of restoration in the presence of noise is not a simple one, however, and 'partial restoring functions' are applied direct to the observed occultation curve rather than to the differential of this.[725] In these partial restorations the restored brightness function is that which would have been observed by a fan beam of Gaussian shape and half-power beamwidth β.

The resolution obtainable by the occultation method is limited by receiver noise, receiver stability, noise interference (which may be reflected from the Moon), and by lunar irregularities. It is not, however, dependent on the wavelength used for the observations, apart from frequency-dependent noise and noise interference. If the signal/noise is taken as n with integration over 1 second of arc, it can be shown that a practical value of beamwidth to be chosen is $\beta \geqslant 50/n^2$.[726] If the signal/noise ratio is 10, the beamwidth may be 0·5 second of arc. It can be seen, therefore, that, if the method is to be employed with weak radio sources, a large collecting area and low-noise receivers are needed.

CHAPTER 8

SENSITIVITY

8.1. The limits of detection

The sensitivity of a radiotelescope is a measure of the weakest source of radio emission that can be detected and is, therefore, directly related to the errors of measurement.

The sensitivity depends on a great number of things: the antenna characteristics, the receiver performance, the time spent on any single observation, the number of pieces of information produced simultaneously (the 'image forming' characteristic), and the size of the region of sky which is to be observed. The signal we want to measure usually appears as a change in the receiver output level as the source passes through the antenna beam. Sometimes we may want to measure variations in the intensity of a variable source. The errors of measurement and limits of detection are set by the fluctuations of the receiver output voltage. These fluctuations are due to:

(*a*) *The signal itself.* The thermal radiation from a black body is governed by statistical laws; the energy flux measured at any point fluctuates in a random fashion about a mean value which is given by Planck's law. Only a fraction of the radiation studied in radio astronomy is thermal in origin, but all the other natural radiation has the same quality of random fluctuations about a mean value called the 'intensity' or 'flux density', which is the object of the measurements. The flux density must, therefore, be defined as the average of the energy flux over a very long—in theory, infinite—time, and normal measurements over seconds or minutes will show the result of insufficient averaging by giving somewhat different answers from one measurement to the next—that is, the output of the receiver will show a greater or smaller fluctuating 'noise' or error component superposed on the wanted mean value.

(*b*) *The medium between the radio source and the telescope.* Clouds of gas in interplanetary space can produce scintillation of sources that have small angular diameters. More troublesome, however, is the Earth's atmosphere, particularly at wavelengths close to the two ends of the available spectrum; variations in the ionosphere produce significant effects near the long wave limit (metre and decimetre waves), while, near the short wave limit (centimetre and millimetre waves), the significant disturbing factor is the attenuation and scattering in the lower atmosphere.

(*c*) *Unwanted signals that reach the telescope.* These are signals from sidelobe directions away from the main beam. Some of them, such as the radiation from the rest of the sky and the thermal emission from the ground, are of the same statistically fluctuating type as the wanted signal but can usually be measured and the necessary corrections can be made by special calibrating procedures. Far more damaging to radio astronomy is man-made radio interference: broadcast, television, radar, etc. The nuisance value of these is illustrated by the great efforts of radio astronomers to get radio frequency channels in which no man-made transmissions are allowed. The interference can be reduced by placing telescopes in remote areas, since the curvature of the Earth and natural barriers such as mountains produce a rapid attenuation of the disturbing signals. This solution, however, is only a partial one, because man-made interference can come from satellites, be reflected from the Moon, or be scattered by irregularities in the atmosphere.

(*d*) *Receiver noise.* The various circuits in the receiver generate unwanted radio noise which adds to that coming from the antenna. This is a major contributor to the total receiver output noise power in the medium and short wave regions of the radio astronomy spectrum; the sky is so 'bright' at the longer wavelengths that the receiver noise becomes less important in comparison. Different types of receivers generate different amounts of noise power and a 'good' receiver should generate as little as possible.

(*e*) *Instabilities in the equipment.* These are troubles described as 'gain variations', 'drifts', 'jumps', etc., which are more common than it would appear from descriptions in published papers. The requirements for stability are often very strict; it is common to measure a source in the presence of more than a hundred times as much unwanted power; a change of less than one per cent in the receiver gain will then produce a false deflection at the receiver output which is equal to that caused by the source we want to measure. One way in which to improve this situation is to modulate the signal in some way before it reaches the receiver and to register only the modulated component of the receiver output. The noise power originating in the receiver, which is often the major contributor, is not so modulated and the stability requirements can be correspondingly relaxed.

8.2. Random noise

The sensitivity of a radiotelescope is usually discussed in terms which suggest that all the sources of error in the measurements are caused by fluctuations with the well-defined statistical properties of black-body radiation. In fact, as we have seen, this is not so: most of the effects described in (*b*), (*c*), and (*e*) are not of this type. However, the sources of noise in this restricted sense lend themselves more easily to mathematical treatment and the resulting sensitivity equations will be valid for ideal observing conditions—no interference, no scintillation, and completely stable equipment. We can in practice often reach close to this ideal performance.

8.2.1. *Bandwidth, integration time and noise fluctuations*

The conditions for optimum reception in radio astronomy differ in many ways from those in radio communication. The radio astronomer is interested in the average energy flux per unit bandwidth but not in the actual structure of the signal fluctuations which, with few exceptions, are due entirely to the statistical nature of the noise radiation. He therefore measures the randomly fluctuating power from the antenna over some period of time which is long enough to give him a sufficiently accurate value of the average power. Simple statistical theory shows that the r.m.s. error in such a determination will be inversely proportional to the square root of the total averaging or *integration time* t sec: the average of N consecutive measurements each lasting, say, 1 second (= one N-second measurement) will be $\sqrt{N}$ times more accurate that a single one-second measurement on its own.

The frequency spectrum of the radiation, i.e. the way in which the energy flux per unit bandwidth depends on the frequency of observation, is another astronomically important parameter, but this information will be lost unless the frequency resolution of the receiver is finer than the spectral details to be explored. This applies to observations of the continuum radiation which covers the whole available frequency spectrum as well as to observations of emission or absorption spectral lines. The power spectrum of the source can, then, usually be regarded as approximately flat over the pass band of the receiver. The r.m.s. error in the measured power per unit bandwidth when averaged over a certain period of time (for example, one second) will be inversely proportional to the square root of the receiver bandwidth $\Delta\nu$ Hz; the average of the measurements taken simultaneously by N receivers working in N adjacent frequency bands (= one measurement with a receiver which has N times the bandwidth) will be $\sqrt{N}$ times more accurate than any one of the N individual measurements on its own. Further, the magnitude of the noise fluctuations is proportional to the average power itself. The r.m.s. error $(\Delta T)_{\text{r.m.s.}}$ as measured with an ideal receiver with a bandwidth $\Delta\nu$ Hz and an integration time t sec is given by

$$(\Delta T)_{\text{r.m.s.}} = \text{const. } T/\sqrt{(\Delta\nu t)}. \qquad (8.1)$$

The noise powers are expressed in terms of noise temperature, the temperature of a resistor from which the same power is available within the same frequency band. The noise temperature is related to the power (expressed in W Hz^{-1}) by Nyquist's formula for the available power from a hot resistor[801]

$$p = kT \text{ watt Hz}^{-1} \qquad (8.2)$$

$k = 1{\cdot}38 \, . \, 10^{-23}$ watt sec (deg K)$^{-1}$. 1° K noise temperature is, therefore, equivalent to $1{\cdot}38 \, . \, 10^{-23}$ watt Hz^{-1} available power.

The strict noise theory, which will not be given here, leads to a value of unity for the proportionality constant in (8.1) in the ideal case of a uniform 'rectangular' passband $\Delta\nu$ Hz wide.[802,803] When the passband has some different shape, or when the time averaging is done in a different way, we should use respectively 'noise equivalent' bandwidth and integration times that have been calculated for some representative instances.[804] The sensitivity equations are often expressed in terms of an RC time constant τ sec and the integration time t in the denominator will then be replaced by the noise equivalent integration time for this circuit, 2τ sec.

Radio astronomy receivers do not, as a rule, behave as ideal square law detectors and the error $(\Delta T)_{\text{r.m.s.}}$ will, in practice, be greater than the value given by (8.1) with constant = 1. We take this into account by introducing a multiplier M which is equal to or larger than unity:

$$(\Delta T)_{\text{r.m.s.}} = M \frac{T}{\sqrt{(\Delta \nu t)}}. \tag{8.3}$$

The exact detector law does not, in fact, make a great deal of difference; a linear detector, for instance, gives $M = 1{\cdot}03$, but there are receiver types for which the multiplier M is much larger than unity.[803] $(\Delta T)_{\text{r.m.s.}}$ is the r.m.s. error in any single t second measurement of the mean noise power.

8.2.2. *Noise fluctuations and gain stability*

The factor $\sqrt{(\Delta \nu t)}$ in (8.3) is often 10^4 or more for radio astronomy receivers and the output fluctuations will represent powers ΔT which are very small compared with the total system noise power $T = T_{\text{sys}}$ °K. The slight change in the output due to a very weak radio source passing through the antenna beam can thus be detected in the presence of a large amount of irrelevant noise coming from the rest of the sky, the ground, and the receiver itself. Such direct measurements require extremely stable equipment: a fractional gain change $\Delta G/G$ will cause an error $(\Delta G/G \,.\, T_{\text{sys}})$ in the measurement of T_{sys}, and gain instabilities rather than statistical noise fluctuations will limit the accuracy if these are greater than one part in $\sqrt{(\Delta \nu t)}$. Such high stabilities are extremely difficult to achieve in practice, and special techniques have been developed to reduce the influence of gain variations on the measurements.

The purpose of 'switched' receivers (Figure 8.1) is to let the receiver measure a small power difference rather than a large total power T_{sys}. The input is switched rapidly between the antenna and a known comparison noise generator, for example, a resistor at a known temperature T_{ref} °K.[802] The resulting modulation of the receiver output power is proportional to the difference $(T_a - T_{\text{ref}})$ between the powers from the antenna (T_a is the antenna temperature) and from the reference generator and can be measured by means of a synchronous detector. The error caused by a gain change $\Delta G/G$ is much smaller in the switched receiver because the modulated component only is measured.

The errors are:

$$\left.\begin{array}{ll}\text{switched receiver} & \dfrac{\Delta G}{G}\,.(T_a - T_{\text{ref}}), \\[2ex] \text{straight receiver} & \dfrac{\Delta G}{G}\,.\,T_{\text{sys}}.\end{array}\right\} \qquad (8.4)$$

Clearly, for optimum stability T_{ref} should be the same as T_a. We can go one step further and adjust T_{ref} continuously to be equal to T_a by a servo loop. The receiver is then used to detect, but not to

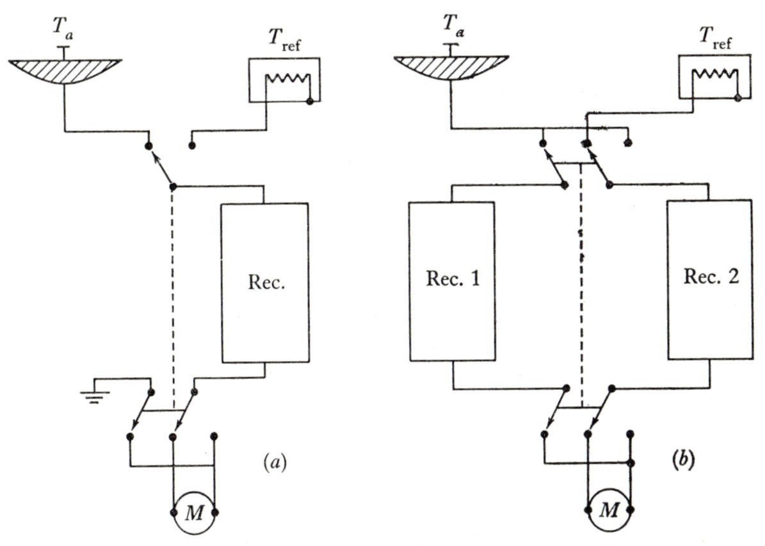

Figure 8.1. Receivers which measure the difference between the output of an antenna and a reference source. (*a*) The receiver spends half the time connected to the antenna and half-connected to the reference source. (*b*) The difference in power is measured continuously by two receivers which are interchanged hundreds of times per second.

measure, power differences, and the gain instabilities should, in principle, have no influence on the measurement.(810)

The increased stability achieved by switching or equivalent techniques is obtained at the expense of an increase in the statistical fluctuations at the receiver output. The receiver spends only part of the time measuring the signal power. If the receiver noise is much greater than the signal, the optimum result is obtained if the time is equally divided between reception of the signal and reception of the comparison source. In a total-power receiver the comparison source is a constant source of known temperature, such as a resistor at constant temperature or else some particular part of the sky. With a phase-switched receiver the comparison is between two different parts of the sky (or the same part differently weighted).[501] The sharing of time between signal and comparison source halves the effective time of observation of the source and increases M by a factor of $\sqrt{2}$.† This $\sqrt{2}$ factor is recovered if two receiving channels (Figure 8.1*b*) are provided and the signal passes alternatively through one or other channel while the comparison signal is connected to the opposite channel.[805] The same effect is obtained in a correlation telescope by direct multiplication (correlation) of the signals from the two antennas, so that switching is not required.[806]

Table 8.1

Telescope	M
Ideal receiver	1
Switched receivers	
One channel, square wave modulation and detection	2
One channel, only fundamental switching frequency detected	2·22
Two channels, square wave modulation and detection	1·41
Two channels, only fundamental switching frequency detected	1·56
Correlation receiver (direct multiplication)	1·41

There is a further factor of $\sqrt{2}$ by which M must be multiplied in practically all radio astronomical measurements. This factor arises from the fact that the measurements are essentially of the difference of two fluctuating voltages—the signal plus receiver noise, and the receiver noise alone. These voltages are usually of similar magnitude, because the signal is small compared with the system noise. If equal times are spent in measuring each of the separate voltages, the fluctuations in each are similar and the fluctuation in the difference is $\sqrt{2}$ times the fluctuation in each. Thus M is increased by a factor of $\sqrt{2}$. In Table 8.1 the values of M are given for various conditions of reception.

† A further increase in M by a factor of $\pi/\sqrt{8}$ occurs if a filter is used to remove the harmonics of the fundamental switching frequency before the final detector in the receiver.

8.2.3. *Noise fluctuations and receiver sensitivity*

The sensitivity of a radio astronomy receiver is usually defined in terms of the minimum detectable change $(\Delta T)_{\min}$ in the input noise power expressed as a temperature. This minimum detectable power must produce an effect larger than any likely chance fluctuation in the receiver output. It has become common practice to define the sensitivity $(\Delta T)_{\min}$ as some multiple of the r.m.s. noise fluctuation, usually $5(\Delta T)_{\text{r.m.s.}}$. The probability of such a large fluctuation happening by chance is only $6 \cdot 10^{-7}$, if, as with a uniform integrator, the fluctuations are distributed according to the normal gaussian error function. The actual distribution does, in fact, depend on the particular integration circuit employed at the receiver output and will generally be less favourable from this point of view. The probability of a change deflection $5(\Delta T)_{\text{r.m.s.}}$, for instance, will be as high as 0·02 when a simple *RC* time constant is used at the output.[807,808] This is an extreme example and not really relevant in practice, because the output from such a receiver is always averaged by hand or in a computer over some period t_1 which is considerably longer than the noise equivalent time $t = 2\tau$, of the *RC* circuit. The effective errors will, then, be more like those of a uniform integrator with an integration time t_1. We can now rewrite (8.3) in terms of the sensitivity $(\Delta T)_{\min}$ and with the noise power T equal to the system noise T_{sys} °K

$$(\Delta T)_{\min} = 5M \frac{T_{\text{sys}}}{\sqrt{(\Delta \nu t)}}. \tag{8.5}$$

The factor 5 is most commonly used to define receiver sensitivity. but others have also been used. The r.m.s. value of the noise fluctuation itself (8.3) is sometimes given as the 'sensitivity', but it is, of course, much smaller than the minimum signal that can be detected with any confidence by a measurement which is completed in t seconds.

8.3. Total-power telescopes

8.3.1. *Noise powers in the system*

The total-power receiver is used to measure the antenna temperature T_a. The system noise power T_{sys}, referred to the receiver input, when the receiver is connected directly to the antenna will be:

$$T_{sys} = T_a + T_N \quad (°K). \tag{8.6}$$

T_N is the *receiver noise temperature* and represents the noise contributions from the receiver itself: the receiver behaves as if there were an extra noise contribution T_N °K at the receiver input. The antenna is usually connected to the receiver via a transmission line with a certain attenuation. When this is taken into account we get a more general expression for T_{sys}:

$$T_{sys} = \eta_t T_a + (1 - \eta_t) T_{290} + T_N \quad (°K), \tag{8.7}$$

η_t is the *efficiency factor of the transmission line* and expresses the fraction of the input power which is available at the other end of the line. Thus $0 \leqslant \eta_t \leqslant 1$ and, if there are no losses, $\eta_t = 1$. The $\eta_t T_a$ term in (8.7) is the available power from the antenna after passing through the transmission line. The second term is the thermal contribution from the transmission line itself. (T_{290} is the real temperature of the material in the line, the so-called 'ambient temperature', usually about 290 °K). The various contributions to T_{sys} are shown in Figure 8.2.

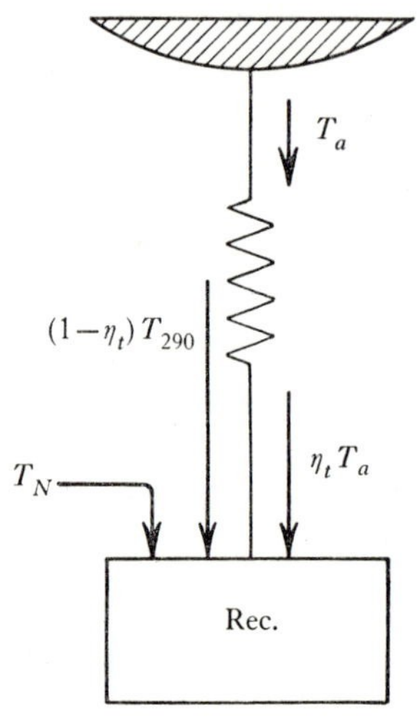

Figure 8.2. The radio frequency power T_a at the output of the antenna is reduced to $\eta_t T_a$ in its passage along the transmission line to the receiver. At the receiver additional thermal noise power $(1-\eta_t) T_{290}$ comes from the transmission line and T_N comes from the input circuits of the receiver itself.

The smallest detectable change in the antenna temperature $(\Delta T_a)_{min}$ must, after suffering a loss η_t in the transmission line, cause a change $(\Delta T)_{min}$ in the system noise temperature, i.e.

$$(\Delta T)_{min} = \eta_t (\Delta T_a)_{min}. \tag{8.8}$$

Substituting in (8.5) and solving for the sensitivity now expressed as the minimum detectable change $(\Delta T_a)_{min}$ we find

$$(\Delta T_a)_{min} = 5M \frac{T_{sys}}{\sqrt{(\Delta \nu t)}\, \eta_t}, \tag{8.9}$$

where $$T_{sys} = \eta_t T_a + (1 - \eta_t) T_{290} + T_N. \tag{8.7}$$

8.3.2. *Noise powers in a system with preamplifiers*

The twofold effect of the transmission line in reducing sensitivity (attenuation of the signal and addition of extra noise) can be serious when the transmission line cannot be made very short, but this situation can be improved considerably by placing a preamplifier close to the antenna terminals in front of the long transmission line to the receiver. The system noise power referred to the input of the *preamplifier* T'_{sys} will be, in analogy with (8.7),

$$T'_{sys} = \eta'_t T_a + (1-\eta'_t)\, T_{290} + T'_N, \tag{8.10}$$

η'_t is the efficiency factor of the short piece of transmission line connecting the antenna output to the preamplifier ($\eta'_t \simeq 1$) and T'_N is the *preamplifier* noise temperature defined in the same way as the receiver noise temperature. The power entering the long transmission line to the receiver will now be $G'T'_{sys}$ instead of T_a (G' is the power gain of the preamplifier) and the system noise temperature referred to the *receiver* input becomes

$$\begin{aligned} T_{sys} &= \eta_t(G'T'_{sys}) + (1-\eta_t)\, T_{290} + T_N \\ &= \eta_t G' \left\{ \eta'_t T_a + (1-\eta'_t)\, T_{290} + T'_N + \frac{(1-\eta_t)\, T_{290} + T_N}{\eta_t G'} \right\}. \end{aligned} \tag{8.11}$$

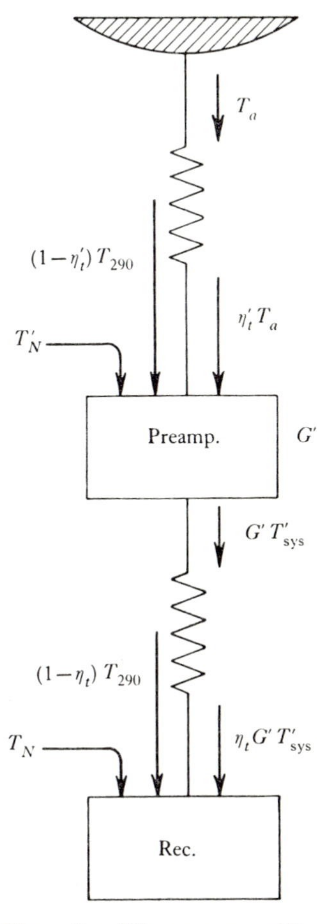

Figure 8.3. When a preamplifier is used, the calculation of the noise reaching the main receiver involves two stages each similar to that in Figure 8.2.

The last term in the bracket, which contains the thermal contribution from the long transmission line and the receiver noise temperature, is reduced in importance in comparison with the other terms by the factor $\eta_t G'$. In particular, if the amplifier gain is very high it can be ignored and

$$T_{\text{sys}} = \eta_t G' T'_{\text{sys}}. \tag{8.12}$$

The least detectable change in antenna temperature $(\Delta T_a)_{\min}$ should produce a change $(\Delta T)_{\min}$ in the system noise temperature T_{sys} given by (8.11). Hence

$$(\Delta T)_{\min} = \eta_t G' \eta'_t (\Delta T_a)_{\min}. \tag{8.13}$$

Substitute (8.12) and (8.13) into (8.5) and solve for $(\Delta T_a)_{\min}$

$$(\Delta T_a)_{\min} = 5M \frac{T'_{\text{sys}}}{\sqrt{(\Delta \nu t)}\, \eta'_t}, \tag{8.14}$$

where
$$T'_{\text{sys}} = \eta'_t T_a + (1 - \eta'_t) T_{290} + T'_N. \tag{8.10}$$

This equation has exactly the same form as (8.9), which therefore applies when a high gain preamplifier is used provided that (1) the (high) efficiency of the short line η'_t replaces the low efficiency of the main transmission line η_t, and (2) the preamplifier noise temperature T'_N replaces the receiver noise temperature T_N.

The approximation (8.12) is not valid when the preamplifier gain is insufficient to eliminate the last term in (8.11) and the exact expression must then be used.

Preamplifiers may be used also in multi-antenna systems. Each antenna has its own preamplifier. For such an arrangement (with high gain preamplifiers) (8.14) will be valid when the preamplifiers and the transmission lines are identical.

8.3.3. *Sensitivity of a total-power telescope with respect to the flux density from a point source*

The matched polarized flux S_m W Hz^{-1} m^{-2} will, by definition, produce an available power $S_m A$ W Hz^{-1} at the antenna output terminals. A m^2 is the effective area in the direction of the point source. Clearly, for optimum result the source should be in the maximum direction, and the maximum contribution to the available power will be $S_m A_{\max}$ W Hz^{-1} or, expressed in noise temperature, $k^{-1} S_m A_{\max}$ °K. The minimum detectable change $(\Delta S_m)_{\min}$ should cause a change $(\Delta T_a)_{\min}$ in the antenna temperature. Thus

$$(\Delta T_a)_{\min} = k^{-1} (\Delta S_m)_{\min} A_{\max} \text{ °K}. \tag{8.15}$$

Substitute this in (8.9) and solve for $(\Delta S_m)_{\min}$

$$(\Delta S_m)_{\min} = 5M \frac{kT_{\text{sys}}}{\sqrt{(\Delta\nu t)}\,\eta_t A_{\max}} \quad (\text{W m}^{-2}\,\text{Hz}^{-1}), \tag{8.16}$$

where
$$T_{\text{sys}} = \eta_t T_a + (1-\eta_t)\,T_{290} + T_N. \tag{8.7}$$

When a high gain preamplifier is used T'_{sys}, η'_t and T'_N take the place of T_{sys}, η_t and T_N respectively. The sensitivity equation can be expressed in terms of the minimum total flux $(\Delta S)_{\min}$ if the polarization of the wave and of the telescope are known; in the simple case of a randomly polarized ('unpolarized') wave we have

$$S_m = S/2 \quad \text{(randomly polarized waves)} \tag{2.3}$$

and (8.16) can be written

$$(\Delta S)_{\min} = 5M \frac{2kT_{\text{sys}}}{\sqrt{(\Delta\nu t)}\,\eta_t A_{\max}} \quad \text{(randomly polarized waves).} \tag{8.17}$$

8.3.4. *Sensitivity of a total-power telescope with respect to the brightness of an extended source*

By 'extended source' we mean a source which occupies a solid angle larger than that of the beam. The measurements can then give (a smoothed version of) its radio brightness distribution. This can be expressed in units of W Hz^{-1} m^{-2} sterad^{-1} (matched polarized brightness $B_m(l, m)$) or in units of °K (matched polarized brightness temperature $T_B(l, m)$). The latter is defined in terms of B_m

$$B_m(l, m) = \frac{k}{\lambda^2}\,T_B(l, m). \tag{2.13}$$

The antenna temperature T_a °K in observations of the sky with a (matched polarization) brightness temperature distribution

$$T_B(l, m)$$

was calculated in Chapter 2 for a lossless antenna:

$$T_a = \lambda^{-2}\int_{4\pi} T_B(l, m)\,A(l, m)\,d\Omega. \tag{2.14}$$

We are interested in the sensitivity to a change at ΔT_B in brightness temperature over the solid angle occupied by the main beam due, for instance, to an extended source entering the beam. The minimum detectable change $(\Delta T_B)_{\text{min}}$ should give rise to a change $(\Delta T_a)_{\text{min}}$ in the antenna temperature. Applying the change $(\Delta T_B)_{\text{min}}$ to the *main beam region* only we derive from (2.14)

$$(\Delta T_a)_{\text{min}} = \lambda^{-2}(\Delta T_B)_{\text{min}} \int_{\text{main beam}} A(l, m)\, d\Omega. \qquad (8.18)$$

The 'main beam region' or 'full beam' is a somewhat vague concept but should contain the parts of the antenna pattern which are directly concerned with the source to be measured, that is, the main beam plus the close sidelobes.[809] The exact definition is not important for the present discussion.

Introduce the *beam efficiency* η_B by

$$\eta_B = \lambda^{-2} \int_{\text{main beam}} A(l, m)\, d\Omega. \qquad (8.19)$$

The beam efficiency η_B includes two factors: (*a*) ohmic losses represented by the radiation efficiency η_R, and (*b*) a factor representing the fraction of the response that occurs in the main beam region.

If there are no far away sidelobe responses, the integral over the main beam region equals that over the whole sky $= \lambda^2 \eta_R$. In practice there are always weak sidelobe responses over the whole sky and the ratio η_B/η_R indicates what fraction of the all-sky integral of $A(l, m)$ is due to the solid angle occupied by the main beam region. The fraction due to the rest of the sky $(1 - \eta_B/\eta_R)$ is often called the *stray factor*. Equation (8.18) can be written:

$$(\Delta T_a)_{\text{min}} = \eta_B (\Delta T_B)_{\text{min}} \qquad (8.20)$$

and (8.9) gives the desired formula for $(\Delta T_B)_{\text{min}}$

$$(\Delta T_B)_{\text{min}} = 5M \frac{T_{\text{sys}}}{\sqrt{(\Delta \nu t)}\, \eta_t \eta_B}, \qquad (8.21)$$

where $$T_{\text{sys}} = \eta_t T_a + (1 - \eta_t)\, T_{290} + T_N \qquad (8.7)$$

and $$\eta_B = \lambda^{-2} \int_{\text{main beam}} A(l, m)\, d\Omega. \qquad (8.19)$$

When a high gain preamplifier is used, the quantities T_{sys}, η_t and T_N should, as usual, be replaced by T'_{sys}, η'_t and T'_N respectively.

8.4. Correlation telescopes

For a *correlation receiver* the sensitivity can be determined in much the same way as for a total-power receiver, but the calculation is less simple because the correlation telescope has two components. We define the correlation sensitivity in a way analogous to the definition of total-power sensitivity. The minimum detectable cosine correlation temperature $(\Delta T_c)_{\min}$ is defined as that which gives a deflection at the cosine receiver output five times greater than the r.m.s. fluctuations produced by the noise in the system. (An analogous definition holds for the minimum detectable sine correlation temperature $(\Delta T_s)_{\min}$ when measurements are made with a sine receiver.) Hence

$$(\Delta T_c)_{\min} = 5M \frac{T_{\text{sys}}}{\sqrt{(\Delta \nu t)}}. \tag{8.22}$$

The definition of system noise T_{sys} has to take into account the possibility that the noise level is not the same in the two channels which are correlated. We define T_{sys} to be the power available from a common output when the two channels are connected together directly (as they are in a phase-switched receiver), even when they are not directly connected. We define the available noise powers in the two channels by the effective temperatures T_1 and T_2 at the junction (Figure 8.4). The available combined power is $\frac{1}{2}(T_1+T_2)$ provided that the powers are uncorrelated. The noise power added after the junction T_N must be added to this. Thus the system temperature is

$$T_{\text{sys}} = \tfrac{1}{2}(T_1+T_2)+T_N, \tag{8.23}$$

where the usually very small correlated component of T_1 and T_2 has been neglected. Thus

$$(\Delta T_c)_{\min} = 5M \frac{\frac{1}{2}(T_1+T_2)+T_N}{\sqrt{(\Delta \nu t)}}. \tag{8.24}$$

If some other definition of T_{sys} is used, the value of M will be different from that given in Table 8.1 for any particular receiving system.

If there were no loss in connecting the antennas to the junction, then T_1 and T_2 would be equal to the antenna temperatures T_{a1} and T_{a2}. Usually there are preamplifiers and transmission losses to be taken into account.

8.4.1. *A correlation system with preamplifiers*

We introduce the system noise temperatures T'_{sys1} and T'_{sys2} referred to the preamplifier inputs, as we did for the total-power telescope. Analogous to (8.7) we have

$$\left.\begin{aligned} T'_{\text{sys1}} &= T_{a1}\eta'_{t1} + (1-\eta'_{t2})\,T_{290} + T'_{N1}, \\ T'_{\text{sys2}} &= T_{a2}\eta'_{t2} + (1-\eta'_{t2})\,T_{290} + T'_{N2}, \end{aligned}\right\} \tag{8.25}$$

where η'_{t1} and η'_{t2} represent the power efficiencies of the transmission lines which connect the antennas to the preamplifiers. T'_{N1} and T'_{N2} are the noise temperatures of the preamplifiers and their power gains are G'_1 and G'_2. The signals pass from the preamplifiers through transmission lines of efficiency η_{t1} and η_{t2} to the receiver. At the receiver inputs the available powers (in terms of temperature) are denoted by T_1 and T_2, where

$$\left.\begin{aligned} T_1 &= T'_{\text{sys1}}\,.\,G'_1\,.\,\eta_{t1} + (1-\eta_{t1})\,T_{290}, \\ T_2 &= T'_{\text{sys2}}\,.\,G'_2\,.\,\eta_{t2} + (1-\eta_{t2})\,T_{290} \end{aligned}\right\} \tag{8.26}$$

and the system noise T_{sys} referred to the correlation receiver input is given by (8.23)

$$T_{\text{sys}} = \tfrac{1}{2}(T_1 + T_2) + T_N. \tag{8.23}$$

If we use this value of T_{sys} in (8.22) we shall get the minimum detectable correlation temperatures at the receiver inputs. However, the sensitivity equations must refer to the signals as measured at the outputs of the two component antennas, i.e. before preamplifiers and transmission lines. The correlation temperature is proportional to the product of the signal voltages and hence to the half power of the products of the signal powers in the two branches. Hence $(\Delta T_c)_{\min}$ is smaller than the minimum detectable values at the receiver inputs by a factor

$$(\eta'_{t1} G'_1 \eta_{t1}\,.\,\eta'_{t2} G'_2 \eta_{t2})^{\frac{1}{2}}.$$

and

$$(\Delta T_c)_{\min} = 5M \frac{\frac{1}{2}(T_1 + T_2) + T_N}{\sqrt{[\Delta\nu t\,.\,(\eta'_{t1} G'_1 \eta_{t1}\,.\,\eta'_{t2} G'_2 \eta_{t2})]}}, \tag{8.27}$$

where T_1 and T_2 are derived from (8.25) and (8.26).

Usually the equations may be simplified, since the preamplifier gains G'_1, G'_2 are so high that noise introduced later into the system

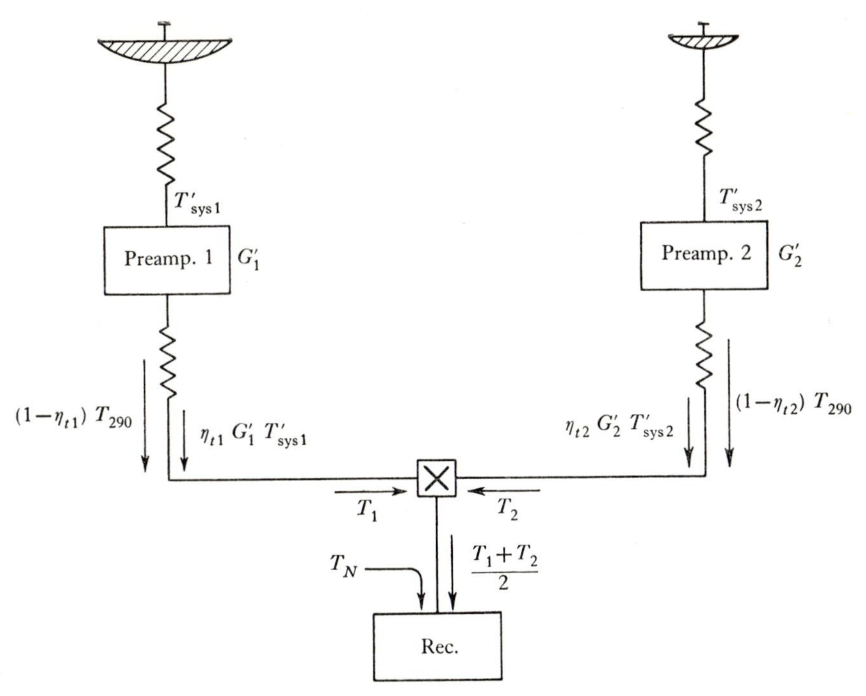

Figure 8.4. The radio-frequency power reaching the receiver of a correlation telescope.

can be ignored. If we put

$$T_1 \approx T'_{\text{sys1}} G'_1 \eta_{t1}, \quad T_2 \approx T'_{\text{sys2}} G'_2 \eta_{t2} \tag{8.28}$$

then (8.28) reduces to $(\Delta T_c)_{\text{min}}$

$$= 5M \frac{1}{\sqrt{(\Delta\nu t)}\sqrt{(\eta'_{t1}\cdot\eta'_{t2})}}\cdot\frac{1}{2}\left[\sqrt{\left(\frac{G'_1\eta_{t1}}{G'_1\eta_{t2}}\right)}\,T'_{\text{sys1}} + \sqrt{\left(\frac{G'_2\eta_{t2}}{G'_2\eta_{t1}}\right)}\,T'_{\text{sys2}}\right]. \tag{8.29}$$

The optimum sensitivity is found when

$$G'_1\eta_{t1}/G'_2\eta_{t2} = T'_{\text{sys2}}/T'_{\text{sys1}} \tag{8.30}$$

that is, when the available powers T_1 and T_2 at the receiver inputs have been adjusted to be equal. Then

$$(\Delta T_c)_{\text{min}} = 5M \frac{\sqrt{(T'_{\text{sys1}}T'_{\text{sys2}})}}{\sqrt{(\Delta\nu t)}\sqrt{(\eta'_{t1}\eta'_{t2})}}, \tag{8.31}$$

where T'_{sys1} and T'_{sys2} are given by (8.28). Equation 8.31 is similar in form to that for a total-power telescope (8.14), with T_{sys} and η'_t

replaced by the geometrical means of the corresponding quantities for the two component antennas. The relations thus derived are for a cosine receiver. For measurements with a sine receiver it is necessary only to replace $(\Delta T_c)_{\min}$ by $(\Delta T_s)_{\min}$ in all equations.

8.4.2. *Sensitivity of a correlation telescope with respect to the flux density from a point source*

The minimum detectable correlation temperature $(\Delta T_c)_{\min}$ is related to the minimum detectable matched polarized flux density $(\Delta S_m)_{\min}$ by

$$k(\Delta T_c)_{\min} = A_c(\Delta S_m)_{\min} \quad (\mathrm{W\,Hz^{-1}}) \tag{8.32}$$

which follows directly from (5.6) and (5.24). We eliminate $(\Delta T_c)_{\min}$ from this equation and (8.31) and solve for $(\Delta S_m)_{\min}$. Then

$$(\Delta S_m)_{\min} = 5M\frac{k\surd(T'_{\mathrm{sys1}}T'_{\mathrm{sys2}})}{\surd(\Delta\nu t)\surd(\eta'_{t1}\eta'_{t2})A_c}. \tag{8.33}$$

Correlation telescopes such as the cross antenna usually have only a cosine receiver and the beam may be steered so that a source can be measured when at the centre of the beam. Then

$$A_c = |\mathbf{A}|_{\max} = 2(A_{1\max}A_{2\max})^{\frac{1}{2}}.$$

Other types of correlation telescopes, particularly interferometers and synthesis telescopes, are often equipped with both a cosine and a sine receiver which are used simultaneously and a source is not necessarily observed at a maximum of the cosine pattern ($\psi = 0$) but in some other direction that corresponds to a phase difference of, say, $\psi = \phi$ at the antenna outputs. If, in front of one receiver input, we introduce a delay which compensates for this phase difference ϕ, we have a telescope whose effective area A_ϕ has an interference maximum in the direction of the source. Then

$$(\Delta S_m)_{\min} = 5M\frac{k\surd(T'_{\mathrm{sys1}}T'_{\mathrm{sys2}})}{\surd(\Delta\nu t)\surd(\eta'_{t1}\eta'_{t2})\,|\mathbf{A}|}, \tag{8.34}$$

where $|\mathbf{A}| = 2(A_1A_2)^{\frac{1}{2}}$ is the value of the envelope pattern in the direction of the source. Now the measurement we would have made with the phasing ϕ, which would include the noise fluctuations of signals from the two branches, can be calculated from the cosine and sine outputs by (5.18)

$$p_\phi = p_c\cos\phi + p_s\sin\phi \tag{5.18}$$

and it follows that (8.34) gives the sensitivity $(\Delta S_m)_{\min}$ for a point source in any direction if the cosine and sine measurements are used in the combination (5.18). Equation (8.33) is seen to be identical with (8.34) in directions where $\psi = 0$ (the sine output due to the source then disappears).

8.4.3. *Sensitivity of a correlation telescope with respect to the brightness temperature of an extended source*

The sensitivity equation for a correlation telescope directed towards an extended source can be derived in a way similar to that for a total-power telescope (section 8.34). Corresponding to (8.18) and (8.19) we have

$$(\Delta T_c)_{\min} = \lambda^{-2}(\Delta T_B)_{\min} \int_{\text{main beam}} A_c(l, m)\, d\Omega, \qquad (8.35)$$

where the subscript c in T_c and $A_c(l, m)$ indicates that the cosine component is under consideration. The *beam efficiency* is defined in the same way as for a filled-aperture telescope (in connection with correlation telescopes the term 'filling factor' is often used for this quantity):

$$\eta_B = \lambda^{-2} \int_{\text{main beam}} A_c(l, m)\, d\Omega \qquad (8.36)$$

and the expression for $(\Delta T_B)_{\min}$ in the correlation telescope becomes

$$(\Delta T_B)_{\min} = 5M \frac{\sqrt{(T'_{\text{sys1}} T'_{\text{sys2}})}}{\sqrt{(\Delta\nu t)}\sqrt{(\eta'_{t1}\eta'_{t2})} \cdot \eta_B}. \qquad (8.37)$$

For some types of correlation telescopes, such as the simple interferometer, $\eta_B \simeq 0$. The instrument does not react to extended sources and (8.37) shows this by producing $(\Delta T_B)_{\min} = \infty$.

8.5. Sensitivity of synthesis telescopes

First, we consider synthesis with a variable spacing interferometer when the movable antenna spends the same time in each element $\delta u \delta v$ of the synthesized aperture and when all these aperture elements are given equal weight (uniform grading of the synthesized aperture). Equation 8.34 for measurements of a point source is then valid without change when t is the *total time of observation* for the whole synthesis: the sensitivity depends only on the value of the envelope pattern in the direction of the source and this does not depend on the relative position of the two antennas on the ground.

In some synthesis systems as, e.g. rotational synthesis (section 7.22) we find that the time δt spent on measurements within equal areas $\delta\sigma$ over the u, v-plane is not constant. The standard sensitivity formulae will be valid if each minute of measurement is given equal weight, i.e. if the synthesized aperture is given the grading $g_s'(u, v)$, where

$$\frac{1}{g_s'} = \text{const.}\,\frac{\delta\sigma}{\delta t}, \tag{8.38}$$

$\delta\sigma/\delta t$ represents the rate at which the synthesis interferometer covers the u, v-plane. If we wish to apply another grading of the synthesized aperture $g_s(u, v) \neq g_s'(u, v)$ we must, in effect, give different weight to equal period measurements taken at different spacings (u, v). This produces a deterioration η_s in the signal/noise and hence the sensitivity over the synthesized map. Error theory gives this as

$$\eta_s = \frac{\Sigma g_s \delta\sigma}{[\Sigma(g_s^2/g_s')\,\delta\sigma\,.\,\Sigma g_s'\delta\sigma]^{\frac{1}{2}}} \tag{8.39}$$

which as expected gives $\eta_s = 1$ for $g_s = g_s'$. As an example we can take a rotational synthesis antenna in which $\delta\sigma/\delta t$ is proportional to r, the east–west spacing between the elements. Then $g_s' = 1/r$ and we may decide to synthesize an aperture of constant weighting $g_s = 1$. Replacing the sums in (8.39) by the corresponding integrals we get $\eta_s = \sqrt{\frac{3}{2}}$, which represents a 13 per cent deterioration in sensitivity when compared either with a system in which the total time of observation has been used in traversing the aperture uniformly, or with a rotational system when used with its 'natural' weighting $g_s = g_s' \propto 1/r$.

The sensitivity equations (8.35) and (8.36) with respect to the brightness temperature of an extended source are, in principle, no different for a synthesis telescope. The definition of the beam efficiency, however, is different, since it must include a weighting loss factor η_s (8.39). The beam over which the effective area is to be integrated is the synthesized beam defined in section 7.1. Since a synthesis telescope is essentially an image-forming device and the sensitivity is not constant over the entire field of the image, we define the sensitivity, to be that at the centre of the image field, i.e. when $\sqrt{(A_1 A_2)}$ is a maximum. Thus (8.37) applies but with η_B given by

$$\eta_B = \lambda^{-2}\eta_s \int_{\text{main synthesized beam}} A_{\text{synth}}(l, m)\, d\Omega. \qquad (8.40)$$

8.6. Image formation and its influence on the effective sensitivity: Definition of surveying sensitivity.

A synthesis telescope is essentially an image-forming instrument, because it measures simultaneously the radiation from an angular region of the sky which is much larger than the synthesized beam of the telescope. Any other type of correlation telescope can also be made to form an image if the telescope is broken up into a number of different parts and the signal from each part is brought separately to a central receiver. With both instruments it is possible to extract as much as N simultaneous pieces of information about the sky brightness temperature, where N is approximately equal to the ratio solid angle beamwidth of the individual antennas to the solid angle beamwidth of the resultant or synthesized beam of the whole telescope.

The ordinary expressions for sensitivity apply only to a small source and do not take into account the size of the field (i.e. the number of pieces of information obtained simultaneously). If the telescope is being used to survey a region very much larger than the angular size of its beam, the image formation is of great advantage. This should obviously be included in a broad definition of sensitivity. A single-beam instrument can use only part of the total available time to observe each beamwidth of the sky, while an image-forming instrument of the same effective area need not divide its time between measurements in different directions.

We define the *surveying sensitivity* $(\Delta SS_m)_{\min}$ as the flux density of the weakest source that can be detected with confidence ($5 \times$ r.m.s. noise error) in a survey of a particular region of sky when the whole survey is completed in a *total time* t_s seconds. We want to express the sensitivity formula in terms of t_s rather than t, the time spent on the measurement in any one particular position of the beam. Let $(\Delta S_m)_{\min}$ be the sensitivity when the telescope is used during the whole time t_s to measure the flux density from one particular direction. If a pencil-beam instrument is required to survey an area Ω_{map} sterad which is larger than the beam Ω_{Beam} sterad we find that

$$t = (\Omega_{\text{Beam}}/\Omega_{\text{map}})\, t_s, \tag{8.41}$$

because the total time t_s must be shared between the $\Omega_{\text{map}}/\Omega_{\text{Beam}}$ different directions. Hence the surveying sensitivity will be

$$(\Delta SS_m)_{\text{min}} = (\Delta S_m)_{\text{min}} \cdot (\Omega_{\text{map}}/\Omega_{\text{Beam}})^{\frac{1}{2}}. \tag{8.42}$$

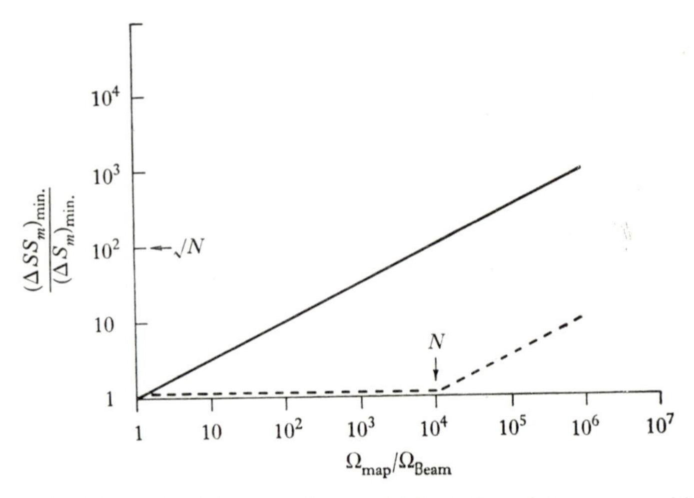

Figure 8.5. The ratio of the surveying sensitivity to the point-source sensitivity for a single response telescope (continuous line) and an image-forming telescope (dotted line) which records N simultaneous pieces of information.

For an image-forming telescope which measures, simultaneously the radiation from all directions within the image $\Omega_i = N\Omega_{\text{Beam}}$ sterad we find that the surveying sensitivity is given by

$$(\Delta SS_m)_{\text{min}} = (\Delta S_m)_{\text{min}}, \tag{8.43}$$

when $\Omega_{\text{map}} \leqslant \Omega_i$

and

$$(\Delta SS_m)_{\text{min}} = (\Delta S_m)_{\text{min}} \sqrt{(\Omega_{\text{map}}/\Omega_i)} = (\Delta S_m)_{\text{min}} \sqrt{(\Omega_{\text{map}}/N\Omega_{\text{Beam}})}, \tag{8.44}$$

when $\Omega_{\text{map}} > \Omega_i$.

A comparison of (8.42) and (8.44) shows that an N-beam image-forming telescope is $\sqrt{N}$ times better in surveying sensitivity than a single-beam instrument when

$$\Omega_{\text{map}} > \Omega_i > N\Omega_{\text{Beam}}$$

and this ratio falls to 1 when $\Omega_{map} = \Omega_{Beam}$. The relationship is shown in Figure 8.5. With the development of antennas of very high resolving power the condition that $\Omega_{map} \gg \Omega_{Beam}$ is becoming the usual one. Hence, the advantage of image-forming over single-beam instruments is becoming ever greater. The role of the single-beam telescope is likely to diminish in future years and its application may be restricted to special problems, as it is with optical telescopes.

APPENDIX 1

In this book directions in the sky are given with respect to the antenna coordinates x, y, z by the directional cosines l, m, n where

$$l = \cos\alpha, \quad m = \cos\beta, \quad n = \cos\gamma. \tag{A 1.1}$$

α, β, and γ are the angles made by the particular direction line with the x, y, and z axes as shown in Figure A 1.1 (We usually specify the x, y plane to be the horizontal plane at the telescope.) Since

$$l^2+m^2+n^2 = 1 \tag{A 1.2}$$

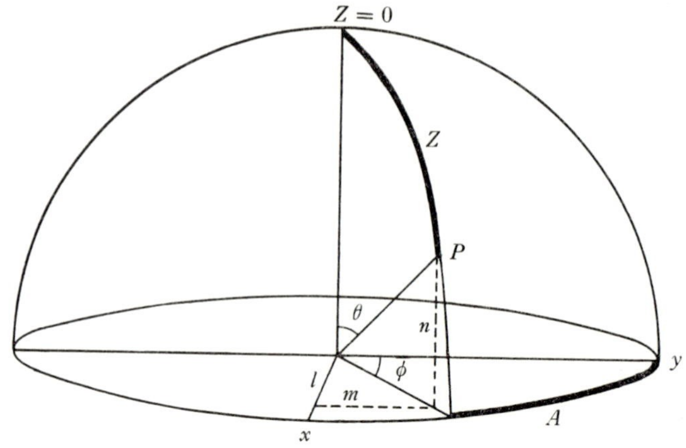

Figure A 1.1. Spherical coordinates and l, m, n.

we need specify only two of l, m, and n. Usually we specify l and m. The value of n which is then derived from (A 1.2) is either positive or negative, but we can usually neglect the negative value, since this is below the horizon.

We can express l, m, and n in terms of θ and ϕ (Figure A 1.1). We measure θ with respect to the z-axis and ϕ with respect to the y-axis; then

$$l = \sin\theta\sin\phi, \quad m = \sin\phi\cos\phi, \quad n = \cos\theta. \tag{A 1.3}$$

In astronomical work θ is usually changed to Z, the zenith distance, and ϕ is measured with respect to either the north point or the south point of the horizon and is called the *azimuth* A. (See Appendix 4.)

APPENDIX 2

2.1. Definition of Fourier transform

If
$$f(l) = \int_{-\infty}^{\infty} g(x) \exp\{j2\pi xl\} dx \tag{A2.1}$$

then
$$g(x) = \int_{-\infty}^{\infty} f(l) \exp\{-j2\pi xl\} dl \tag{A2.2}$$

which we write
$$f(l) \mathscr{F} g(x) \tag{A2.3}$$

$$g(x) \mathscr{F} f(l). \tag{A2.4}$$

2.2. Even and odd functions

A function $g(x)$ may have even and odd parts, i.e.

$$g(x) = \text{even}\, g(x) + \text{odd}\, g(x). \tag{A2.5}$$

If $g(x)$ is also complex, then

$$g(x) = \text{Re}\, g(x) + j\, \text{Im}\, g(x). \tag{A2.6}$$

Now
$$\text{even}\ g(x) = \tfrac{1}{2}[g(x) + g^*(-x)] \tag{A2.7}$$

$$\text{odd}\ g(x) = \tfrac{1}{2}[g(x) - g^*(-x)] \tag{A2.8}$$

and
$$\text{Re}\, g(x) = \tfrac{1}{2}[g(x) + g^*(x)], \tag{A2.9}$$

$$j\, \text{Im}\, g(x) = \tfrac{1}{2}[g(x) - g^*(x)]. \tag{A2.10}$$

If
$$f(l) \mathscr{F} g(x), \tag{A2.3}$$

then
$$\text{Re} f(l) \mathscr{F} \text{even}\, g(x) \tag{A2.11}$$

$$j\, \text{Im} f(l) \mathscr{F} \text{odd}\, g(x). \tag{A2.12}$$

If we separate the real and imaginary parts of $g(x)$, then

$$\text{even Re} f(l) \mathscr{F} \text{even Re}\, g(x) \tag{A2.13}$$

$$\text{even}\, j\, \text{Im} f(l) \mathscr{F} \text{even}\, j\, \text{Im}\, g(x) \tag{A2.14}$$

$$\text{odd}\, j\, \text{Im} f(l) \mathscr{F} \text{odd Re}\, g(x) \tag{A2.15}$$

$$\text{odd Re} f(l) \mathscr{F} \text{odd}\, j\, \text{Im}\, g(x) \tag{A2.16}$$

A Hermitian function is one in which the real part is even and the imaginary part is odd, i.e. $g(x) = g^*(-x)$. The Fourier transform of such a function is real and, conversely, the transforms of all real functions are Hermitian. The term 'even' is often used instead of 'Hermitian' for such a complex function and we follow this custom here.

2.3 Some theorems

Some of the more important properties of Fourier transforms are listed below:

(*a*) The *addition theorem* states that the Fourier transforms of the sum of several functions is the sum of their individual transforms, i.e.

$$\sum_i f_i(l) \mathcal{F} \sum_i g_i(x). \tag{A2.17}$$

This follows directly from the definition of a Fourier transform, since the integral of a sum equals the sum of the individual integrals.

(*b*) The *similarity theorem* states that, if $f(l) \mathcal{F} g(x)$, then, the transform of $g(ax)$ is as follows:

$$\frac{1}{|a|} f\left(\frac{l}{a}\right) \mathcal{F} g(ax) \tag{A2.18}$$

which implies that, as the scale of one function is widened, the scale of its transform is narrowed (e.g. the wider is the aperture of a telescope, the narrower is its angular response).

Equation A2.18 is derived simply. The Fourier transform of $g(ax)$ is, by definition,

$$\int_{-\infty}^{\infty} g(ax) . \exp[j2\pi . lx]\, dx$$

$$= \frac{1}{|a|} \int_{-\infty}^{\infty} g(ax)\, exp\left[j2\pi \frac{l}{a} . ax\right] d(ax)$$

$$= \frac{1}{|a|} \cdot f\left(\frac{l}{a}\right). \tag{A2.19}$$

(*c*) The *shift theorem* states that, if a linear phase shift $\exp(j2\pi sx)$ is introduced in $g(x)$ so that the new function of x is

$$g(x) . \exp(j2\pi sx),$$

then its Fourier transform changes from $f(l)$ to $f(l+s)$. (See (2.57) in Chapter 2.)

(*d*) *Parseval's or Rayleigh's theorem* states that the integrals of the squares of the module of any function and its Fourier transform are equal, i.e.

$$\int_{-\infty}^{\infty} |f(l)|^2 \, dl = \int_{-\infty}^{\infty} |g(x)|^2 \, dx. \tag{A2.20}$$

This is proved by writing

$$|f(l)|^2 = f(l) \cdot f^*(l),$$

replacing the last factor by its Fourier integral and changing the order of integration.

(*e*) The *convolution theorem* was discussed in Chapter 2. The convolution of two functions is defined as

$$g_1(x) * g_2(x) \equiv \int g_1(\sigma) g_2(x-\sigma) \, d\sigma \tag{2.54}$$

and the theorem states that, if $f_1(l)$ and $f_2(l)$ are, respectively, the Fourier transforms of $g_1(x)$ and $g_2(x)$, then the Fourier transforms of the convolution integral is given by

$$f_1(l) \cdot f_2(l) \,\mathcal{F}\, g_1(x) * g_2(x). \tag{2.55}$$

Conversely,

$$f_1(l) * f_2(l) \,\mathcal{F}\, g_1(x) \cdot g_2(x) \tag{2.56}$$

To prove (2.56) we consider the Fourier transform of the product $g_1(x) g_2(x)$ which, by definition, is

$$\int_{-\infty}^{\infty} g_1(x) \cdot g_2(x) \cdot \exp(j2\pi lx) \, dx. \tag{A2.21}$$

We remember that $g_1(x) \,\mathcal{F}\, f_1(l)$, i.e. (replacing l by σ)

$$g_1(x) = \int_{-\infty}^{\infty} f_1(\sigma) \cdot \exp[-j2\pi x\sigma] \, d\sigma. \tag{A2.22}$$

If we substitute A2.22 in A2.21 we have

$$\begin{aligned}
\int_{-\infty}^{\infty} & g_1(x) g_2(x) \exp(j2\pi lx) \, dx \\
&= \int_{-\infty}^{\infty} f_1(\sigma) \int_{-\infty}^{\infty} g_2(x) \exp[j2\pi x(l-\sigma)] \, dx \, d\sigma \\
&= \int_{-\infty}^{\infty} f_1(\sigma) \cdot f_2(l-\sigma) \, d\sigma \\
&= f_1(l) * f_2(l).
\end{aligned} \tag{A2.23}$$

(*f*) *Smoothing and convolution.* For convolution we have

$$g_1(x) * g_2(x) \equiv g_2(x) * g_1(x). \tag{A 2.24}$$

But for the smoothing of two functions

$$\begin{aligned}\int_{-\infty}^{\infty} g_1(\sigma)\, g_2(\sigma - x)\, d\sigma &\equiv g_1(x) \star g_2(x) \quad \text{by definition} \\ &\equiv g_1(x) * g_2(-x) \\ &\equiv g_2(-x) * g_1(x) \\ &\equiv g_2(-x) \star g_1(-x). \end{aligned} \tag{A 2.25}$$

(See Figures 2.8 and 5.3.)

Observe that if

$$f_1(l) \mathscr{F} g_1(x) \quad \text{and} \quad f_2(l) \mathscr{F} g_2(x)$$

then

$$f_1(l) \,.\, f_2(l) \mathscr{F} g_1(x) \star g_2(x).$$

APPENDIX 3

The *available power* from an antenna is defined as the power dissipated in the terminal load when the impedances have been adjusted for maximum transfer of energy from antenna to load—that is, when the load impedance is matched to the antenna impedance.

When the two component antennas of a correlation telescope are used, separately, as total-power telescopes to measure a particular point source, the available powers w_1 and w_2 due to the received radiation from this source will be

$$w_1 = \tfrac{1}{4}\overline{V_{c1}^2}/R_1, \quad w_2 = \tfrac{1}{4}\overline{V_{c2}^2}/R_2, \quad (\mathrm{W_{att}}) \qquad (\mathrm{A}\,3.1)$$

R_1 and R_2 are the internal impedances of the two antennas respectively and V_{c1} and V_{c2} the components of their open circuit voltages that are due to the radiation from the point source we are measuring.

The *available correlated power* from the two antennas when these are combined into a correlation telescope may be defined similarly. We showed in section 5.1 that the correlated power could be expressed formally as the difference between the available powers when the two antennas are joined together directly and when joined in phase reversal. We define the available correlated power from the correlation telescope as this difference when the impedances have been adjusted so as to make this power difference as large as possible.

The open circuit voltage V at the junction after the two antennas have been connected together is

$$V = \frac{R_1 V_{c1} \pm R_2 V_{c2}}{R_1 + R_2} \quad (\mathrm{V_{olt}}), \qquad (\mathrm{A}\,3.2)$$

where the minus sign refers to the case when the signal from the second antenna has been reversed in phase before the junction.

The available power w (W) from the junction is given by

$$w = \tfrac{1}{4}\overline{V^2}/R \quad (\mathrm{W_{att}}), \qquad (\mathrm{A}\,3.3)$$

where R is the internal impedance of the antenna combination measured at the junction and given by

$$1/R = 1/R_1 + 1/R_2 \quad (\mathrm{ohm}^{-1}). \qquad (\mathrm{A}\,3.4)$$

The difference in available power in the two conditions is then

$$\Delta w = \frac{1}{4R}\overline{\left(\frac{R_1 V_{c1} + R_2 V_{c2}}{R_1 + R_2}\right)^2} - \frac{1}{4R}\overline{\left(\frac{R_1 V_{c1} - R_2 V_{c2}}{R_1 + R_2}\right)^2}. \qquad (A\,3.5)$$

Substituting the expression for $1/R$, we find

$$\Delta w = \overline{V_{c1} V_{c2}}/(R_1 + R_2). \qquad (A\,3.6)$$

For narrow band noise voltages we have

$$\begin{aligned} \overline{V_{c1} V_{c2}} &= (\overline{V_{c1}^2}\,\overline{V_{c2}^2})^{\frac{1}{2}} \cos \Psi \\ &= 4(R_1 R_2)^{\frac{1}{2}} (w_1 w_2)^{\frac{1}{2}} \cos \Psi. \end{aligned} \qquad (5.7)$$

The next step follows from A 3.1. Equation A 3.6 becomes

$$\Delta w = \frac{4(R_1 R_2)^{\frac{1}{2}}}{R_1 + R_2} (w_1 w_2)^{\frac{1}{2}} \cos \Psi. \qquad (A\,3.7)$$

The maximum power difference occurs when the antenna impedances are adjusted to be equal, i.e. when $R_1 = R_2$. The first factor on the R.H.S. of A 3.7 is now equal to 2, and the equation, in terms of available power for unit bandwidth becomes

$$p_c = 2(p_1 p_2)^{\frac{1}{2}} \cos \Psi \quad (\mathrm{W\,Hz^{-1}}) \qquad (A\,3.8)$$

which is identical with (5.11), section 5.22.

APPENDIX 4

In Figure A4.1 celestial angular coordinates of a point P in space are shown. One set of coordinates is based on the direction of the Earth's axis. The angle that a star–Earth line makes with the equatorial plane of the Earth is called the *declination* (δ or Dec.),

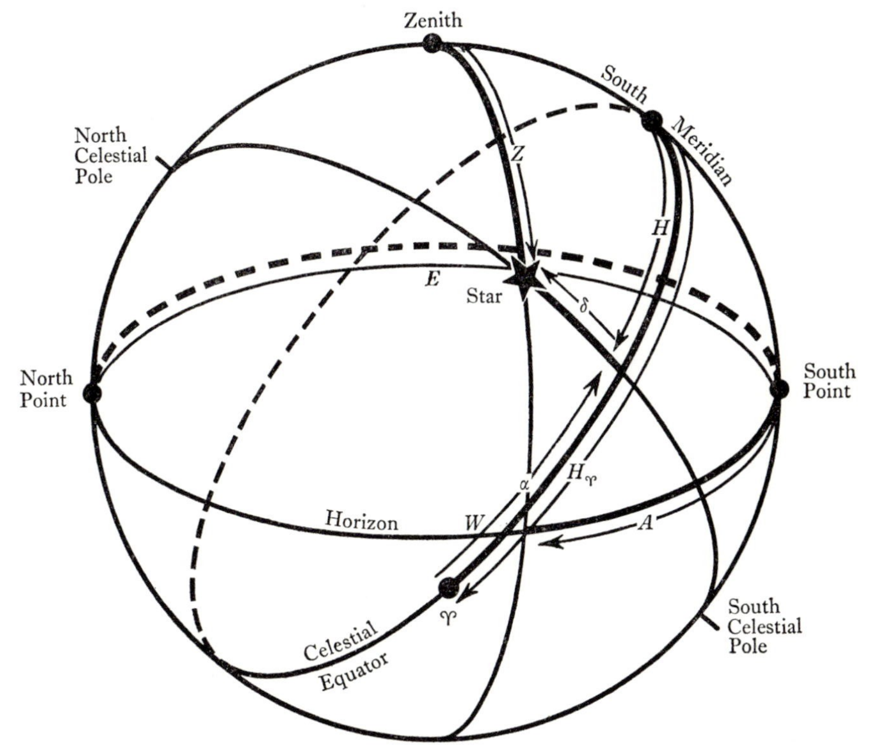

Figure A4.1. Equatorial and alt-azimuth coordinates.

The second celestial coordinate is the angle that the star–Earth line makes with a plane in which the Earth's axis lies and which passes through a particular direction in the sky called the first point of Aries (♈). This is the position of the Sun at the March Equinox. This angle is called the right-ascension (α or R.A.) of the star. If, instead of being measured with respect to this plane, the angle is measured with respect to the meridian plane through the place at which observations are made, the angle is called the *hour-angle* H. The right ascension of a star, therefore, is the difference of the hour–angles of the first point in Aries and of the star, i.e.

$$\alpha = H_{\Upsilon} - H = \Theta - H. \tag{A4.1}$$

At any place on the Earth H_{Υ} is equal to the *sidereal time* Θ at that place, since Θ is measured from the meridian passage of the first point of Aries. Usually α, Θ, and H are expressed in sidereal hours, minutes and seconds rather than in degrees.

Unfortunately, the direction of the Earth's axis is not fixed in space but precesses. This introduces changes in a year of up to 20 seconds of arc in the declination δ of a star and changes of a few seconds of time in its right ascension α. To overcome this difficulty the angles are corrected to give the position at some particular epoch, e.g. 1950 or 2000 A.D The change in position of a star from one epoch to the next is approximately

$$\delta_n - \delta_0 = 20''{\cdot}0 \cos\alpha_0 \,.\, n, \tag{A4.2}$$

where n is the number of years since the epoch at which $\delta = \delta_0$. The change in right ascension is (in seconds of time)

$$\alpha_n - \alpha_0 = 3^{\mathrm{s}}{\cdot}07 + 1^{\mathrm{s}}{\cdot}34 \sin\alpha_0 \,.\, \tan\delta_0. \tag{A4.3}$$

The second set of cordinates is based on the directions of the horizon and of the north–south line on the Earth at the place at which observations are made. The coordinates of a star are measured in this system by its altitude (or, more usually, by the complement of this, the zenith angle Z), and by the azimuth A with respect to the meridian plane through the point of observation.

To convert the position of a star from equatorial to altitude-azimuth coordinates we use the following trigonometrical relations:

$$\cos Z = \cos\delta \cos H \cos\phi + \sin\delta \sin\phi, \tag{A4.4}$$

where ϕ is the terrestrial latitude of the observatory.

The azimuth angle A of the star is then found:

$$\cos A = (\sin\delta - \sin\phi \cos \mathrm{Z})/(\cos\phi \sin Z). \tag{A4.5}$$

In this book directions in the sky have been expressed in terms of the direction cosines l, m, n of the angles α, β, γ. α, β, γ are the angles with respect to the axes x, y, z. The relations between l, m, n and the commonly used astronomical coordinates are now given.

It is convenient to take the x, y plane to be the horizontal plane and to point the x, y, z axes in the west, south and zenith directions respectively. Then if A is measured from the *south* point of the horizon

$$l = \sin Z \sin A, \quad m = \sin Z \cos A, \quad n = \cos Z. \tag{A 4.6}$$

For the equatorial coordinates

$$\left.\begin{aligned} l &= \cos\delta \sin H, \\ m &= \cos\delta \cos H \sin\phi - \sin\delta \cos\phi, \\ n &= \cos\delta \cos H \cos\phi + \sin\delta \sin\phi. \end{aligned}\right\} \tag{A 4.7}$$

Rotational Synthesis. For this important special case we choose the x, y plane perpendicular to the rotational axis of the earth. We may make the directions of the axes as follows:

$$\begin{aligned} &x \text{ axis:} \quad \alpha = \alpha_c - 6 \text{ hours}, \quad && \delta = 0, \\ &y \text{ axis:} \quad \alpha = \alpha_c, && \delta = 0, \\ &z \text{ axis:} && \delta = +90^\circ, \end{aligned}$$

where α_c is the R.A. of the centre of the image field.

The relations between the l, m, n coordinates and the equatorial coordinates are then

$$\left.\begin{aligned} l &= -\cos\delta \sin(\alpha - \alpha_c), \\ m &= \cos\delta \cos(\alpha - \alpha_c), \\ (n &= \sin\delta) \end{aligned}\right\} \tag{A 4.8}$$

and

$$\tan\delta = n/(l^2 + m^2)^{\frac{1}{2}}, \quad \tan(\alpha - \alpha_c) = l/m. \tag{A 4.9}$$

REFERENCES

GENERAL

001 Pawsey, J. L. and Bracewell, R. N. (1954). *Radio astronomy*. Oxford.
002 Steinberg, J.-L. and Lequeux, J. (1960). *Radioastronomie*. Paris: Dunod.
003 Brown, R. H. and Lovell, A. C. B. (1958). *Exploration of space by radio*. New York: Wiley.
004 Smith, F. G. (1960). *Radio astronomy*. Penguin.
005 Kraus, J. D. (1966). *Radio astronomy*. New York: McGraw-Hill.
006 Skobel'tsyn, D. V. (ed.) (1966). *Proc. (Trudy) P. N. Lebedev Phys. Inst. Moscow*, **28**.
007 Bracewell, R. N. (1962). *Handbuch der Physik*, **54**.
008 Ko, H. C. (1964). Radio telescope antennas, in *Microwave scanning antennas*. H. C. Hansen (ed.). New York: Academic Press.
009 *Progress in Radio Science 1963–66*. (1967). Part II: XV, General Assembly of U.R.S.I., Commission V. J. W. Findlay (ed.).

CHAPTER 1

101 Reber, G. (1944). *Astrophys. J.* **100**, 279.

CHAPTER 2

201 Schelkunoff, S. A. (1943). *Electromagnetic waves*. New York: Van Nostrand.
202 Silver, S. (1949). *Microwave antenna theory and design*. New York: McGraw-Hill.
203 Arsac, J. (1961). *Transformation de Fourier et théorie des distributions*. Paris: Dunod.
204 Jennison, R. (1961). *Fourier transforms and convolutions for the experimentalist*. Oxford: Pergamon.
205 Bracewell, R. N. (1965). *The Fourier Transform and its applications*. New York: McGraw-Hill.

CHAPTER 3

301 Phillips, C. J. E (1967). Ph.D. Thesis, University of Sydney.
302 Ruze, J. (1952). *Nuovo Cimento (Suppl.)*, **9**, 364.
303 Robieux, J. (1966). *Ann. Radioélectr.* **11**, 633.
304 Ruze, J. (1966). *I.E.E.E.Proc.* **54**, 633.
305 Minnett, H. C. and Yabsley, D. E. (1966). *I.R.E.E. (Aust.) Proc.* **27**, 304.
306 Puttock, M. J. and Minnett, H. C. (1966). *I.E.E. Proc.* **113**, 1723.
307 Booker, H. G. (1947). *I.E.E. Journ.* (III), **94**, 171.

308 Kaplun, V. A., Babkin, N. I. and Goryachev, B. G. (1964). *Radiotechnica i Electronika* (Eng. Trans.), **9**, 1428.
309 Booker, H. G. (1946). *I.E.E. Journ.* (III A), **93**, 620.
310 Pawsey, J. L. and Harting, E. (1960). *Aust. J. Phys.* **13**, 740.
311 Wielebinski, R. and Shakeshaft, J. R. (1962). *Nature, Lond.* **195**, 982.
312 Kelleher, K. S. and Coleman, H. P., *U.S. Naval Res. Lab. Rept.* 4088.
313 Cheo, B. R. S., Rumsey, V. H. and Welch, W. J. (1961). *I.R.E. Trans. AP*, **9**, 527.
314 Du Hamel, R. H. and Isbell, D. E. (1957). *I.R.E. Convention Record (P.T.I.)*, 119.
315 Dyson, J. D. (1959). *I.R.E. Trans., AP*, **7**, 329.
316 Struve, O., Emberson, R. M. and Findlay, J. W. (1960). *Pub. Astron. Soc. Pacific*, **72**, 439.
317 McAlister, K. R. and Labrum, N. R. (1967). *I.R.E.E. Aust. Proc.* **28**, 291.
318 Husband, H. C. (1958). *Inst. Civil Engrs. London Proc.* **9**, 65; **10**, 410.
319 Lovell, A. C. B. (1957). *Nature, Lond.* **180**, 60.
320 Bowen, E. G. and Minnett, H. C. (1963). *I.R.E. Aust. Proc.* **24**, 98.
321 Wallis, B. (1955). *British Patent App.* 29248/55.

CHAPTER 4

401 Mills, B. Y. and Little, A. G. (1953). *Aust. J. Phys.* **6**, 272.
402 Sichak, W. and Levine, D. J. (1955). *I.R.E. Proc.* **43**, 1661.
403 Swenson, J. R. and Lo, Y. T. (1961). *I.R.E. Trans. AP*, **9**, 9.
404 Hannan, P. W. (1967). *Radio Science*, **2**, 361.
405 Laffineur, M. and Coupiac, P. (1967). *Bulletin de la Société Royale de Liège*, **36**, 393.
406 Swarup, G. (1965). *9th Sympos. Cosmic Rays, Elem. Particle Phys. and Astrophys, Bombay*. (1966) *Proc.* 745.
407 Ryle, M. (1960). *I.E.E. Journ.* **6**, 14.
408 Christiansen, W. N. and Högbom, J. A. (1961). *Nature, Lond.* **191**, 215.
409 Braccesi, A. and Ceccarelli, M. (1962). *Nuovo Cimento*, **23**, 208.
410 Kraus, J. D. (1958). *I.R.E. Proc.* **46**, 92.
411 Boischot, A., Ginat, M. and Parise, M. (1964). *Notes et Informations de l'Observatoire de Paris*, no. 21.
412 Ginat, M. and Steinberg, J.-L. (1967). *Rev. de Phys. App.* **2**, 79.
413 Khaikin, S. E. and Kaidanovsky, N. L. (1959). *Priroda Tech. Expt.* **2**, 19.
414 Chu, L. J. (1961). *Astrophys. J.* **134**, 927.
415 Ashmead, J. and Pippard, A. B. (1946). *I.E.E. Journ.* (III A), **93**, 627.
416 Head, A. K. (1957). *Nature, Lond.* **179**, 692.
417 Geruni, P. M. (1964). *Radiotechnica i Electronika* (Eng. Trans.), **9**, 1.
418 Spencer, R. A., Sletten, C. J. and Walsh, J. E. (1950). *Proc. Nat. Electron. Conf.* **5**, 320.
419 Love, A. W. (1962). *I.R.E. Trans. AP*, **10**, 527.

CHAPTER 5

501 Ryle, M. (1952). *Roy. Soc. Proc.* A, **211**, 351.
502 Mills, B. Y. and Little, A. G. (1953). *Aust. J. Phys.* **6**, 272.
503 Bracewell, R. N. and Roberts, J. A. (1954). *Aust. J. Phys.* **7**, 615.
504 Bracewell, R. N. (1962). *Handbuch der Physik*, **54**, 42.

CHAPTER 6

601 Christiansen, W. N. (1953). *Nature, Lond.* **171**, 831.
602 Covington, A. E. and Broten, N. W. (1957). *I.R.E. Trans. AP* **5**, 247.
603 Bracewell, R. N. and Swarup, G. (1961). *I.R.E. Trans. AP* **9**, 22.
604 Tanaka, H. and Kakinuma, T. (1963). *Proc. Res. Inst. Atmospherics Nagoya Univ.* **10**, 25.
605 Christiansen, W. N. and Wellington, K. J. (1966). *Nature, Lond.* **209**, 1173.
606 Covington, A. E., Legg, T. H. and Bell, M. B. (1967). *Solar Physics*, **1**, 465.
607 Vitkevich, V. V. (1961). *Vestn. Akad. Nauk USSR*, **5**, 23.
608 Mills, B. Y., Aitchison, R. E., Little, A. G. and McAdam, W. B., (1963). *I.R.E. Aust. Proc.* **24**, 156.
609 Christiansen, W. N. and Mathewson, D. S. (1958). *I.R.E. Proc.* **46**, 127.
610 Christiansen, W. N., Erickson, W. C. and Högbom, J. A. (1963). *I.R.E. Aust. Proc.* **24**, 219.
611 Wild, J. P. (1965). *Roy. Soc. Proc.* A **286**, 499.
612 Blum, E. J. (1961). *Ann d'Astrophys.* **24**, 359.

CHAPTER 7

701 McCready, L. L., Pawsey, J. L. and Payne-Scott, R. (1947). *Roy. Soc. Proc.* A **190**, 357.
702 Stanier, H. M. (1950). *Nature, Lond.* **165**, 354.
703 Lequeux, J., Le Roux, E. and Vinokur, M. (1959). *Comptes Rendus (Paris)*, **249**, 634.
704 Read, R. R. (1960). *Obs. Cal. Inst. Tech. Radio Observ.*
705 Blythe, J. H. (1957). *Mon. Not. Roy. Astron. Soc.* **117**, 644.
706 Costain, C. H. and Smith, F. G. (1960). *Mon. Not. Roy. Astron. Soc.* **121**, 405.
707 Scott, P. F., Ryle, M. and Hewish, A. (1961). *Mon. Not. Roy. Astron. Soc.* **122**, 95.
708 Christiansen, W. N. and Warburton, J. A. (1955). *Aust. J. Phys.* **8**, 474.
709 Cooley, J. W. and Tukey, J. W. (1965). *Math. of Comput.* **19**, 297.
710 Jansky, K. G. (1933). *I.R.E. Proc.* **21**, 1387.
711 Bolton, J. G. and Stanley, G. J. (1948). *Aust. J. Sc. Res.* A **1**, 58.
712 Ryle, M. (1962). *Nature, Lond.* **194**, 517.

713 Muller, C. A. (1967). *I.A.U. XIII Assembly Draft Report*, 905.
714 Blum, E. J., Delannoy, J. and Joshi, M. (1960). *Comptes Rendus (Paris)*, **252**, 2517.
715 Morimoto, M. and Labrum, N. R. (1967). *I.R.E. Aust. Proc.* **28**, 352.
716 Goddard, B. R., Watkinson, A. and Mills, B. Y. (1960). *Aust. J. Phys.* **13**, 665.
717 Brown, R. H., Palmer, H. P. and Thompson, A. R. (1955). *Phil. Mag.*, **46**, 857.
718 Adgie, R. L., Gent, A., Slee, O. B., Frost, A. D., Palmer, H. P., and Rowson, B. (1965). *Nature, Lond.* **208**, 275.
719 Broten, N. W. *et al.* (1967). *Science* **156**, 1592.
720 Moran, J. M. *et al.* (1967). *Science* **157**, 676.
721 Bare, C. *et al.* (1967). *Science* **157**, 189.
722 Brown, R. H. and Twiss, R. W. (1954). *Phil. Mag.* **45**, 663.
723 Getzmanzev, G. G. and Ginsberg, W. L. (1950). *J. Expt. and Theor. Phys.* (*USSR*), **20**, 347.
724 Hazard, C., Mackey, M. B. and Shimmins, A. J. (1963). *Nature*, **197**, 1037.
725 Scheuer, P. A. G. (1962). *Aust. J. Phys.* **15**, 333.
726 von Hoerner, S. (1964). *Astrophys. J.* **140**, 65.

CHAPTER 8

801 Nyquist, H. (1928). *Phys. Rev.* **32**, 110.
802 Dicke, R. H. (1946). *Rev. Sc. Inst.* **17**, 268.
803 Lampard, D. G. (1954). *I.E.E. Proc.* (iv), **101**, 118.
804 Colvin, R. S. (1961). *Stanford Electron. Lab. Radio Sc. Lab. Report 18*. Stanford.
805 Muller, C. A. (1956). *Philips Tech. Rev.* **17**, 351.
806 Blum, E. J. (1959). *Ann. d'Astrophys.* **22**, 140.
807 Strum, P. D. (1958). *I.R.E. Proc.* **46**, 43.
808 Robinson, B. J. (1964). *Ann. Rev. Astron. and Astrophys.* **2**, 401.
809 Seeger, C. L., Westerhout, G. and van de Hulst, H. C. (1956). *Bull. Astron. Instit. Netherlands* **13**, 89.
810 Ryle, M. and Vonberg, D. D. (1948). *Roy. Soc. Proc.* A, **193**, 98.

INDEX